SELFING and CROSSING CONCEPT

SELFING and CROSSING CONCEPT

Applied to the Strength of Blended Concrete Mixes

Nisith K. Bairagi
B.C.E., M.E. (Civil), Ph.D.
Retired Professor of Structural Engg.
IIT-Bombay

PUBLISHING FOR ONE WORLD

NEW AGE INTERNATIONAL (P) LIMITED, PUBLISHERS
New Delhi • Bangalore • Chennai • Cochin • Guwahati • Hyderabad
Jalandhar • Kolkata • Lucknow • Mumbai • Ranchi
Visit us at **www.newagepublishers.com**

Published by New Age International (P) Ltd., Publishers
First Edition: 2009

Branches:

- 36, Malikarjuna Temple Street, Opp. ICWA, Basavanagudi, **Bangalore**. © (080) 26677815
- 26, Damodaran Street, T. Nagar, **Chennai**. © (044) 24353401
- Hemsen Complex, Mohd. Shah Road, Paltan Bazar, Near Starline Hotel, **Guwahati**. © (0361) 2543669
- No. 105, 1st Floor, Madhiray Kaveri Tower, 3-2-19, Azam Jahi Road, Nimboliadda, **Hyderabad**. © (040) 24652456
- RDB Chambers (Formerly Lotus Cinema) 106A,1st Floor, S.N. Banerjee Road, **Kolkata**. © (033) 22275247
- 16-A, Jopling Road, **Lucknow**. © (0522) 2209578
- 142C, Victor House, Ground Floor, N.M. Joshi Marg, Lower Parel, **Mumbai**. © (022) 24927869
- 22, Golden House, Daryaganj, **New Delhi**. © (011) 23262370, 23262368

ISBN: 978-81-224-2581-9

Rs. 495.00

C-09-04-3518

Printed in India at Glorious Printers, Delhi.
Typeset at Goswami Associates, Delhi.

PUBLISHING FOR ONE WORLD
NEW AGE INTERNATIONAL (P) LIMITED, PUBLISHERS
4835/24, Ansari Road, Daryaganj, New Delhi-110002
Visit us at **www.newagepublishers.com**

Dedication

This Book is dedicated to those who

Boost Research Inspiring Technical Application

Preface

Concrete or sand-cement mortars develop strength in the presence of water when added to the dry mixture of the ingredients. The development of strength in the sand mass starts from the instant of time t when water is just added to it ($t = 0$). Up to a certain time, preferably the initial setting time t_i , the prepared mass remains plastic and can be moulded easily, after which the hardening process proceeds faster, and towards the final setting time t_f , the mass becomes hard and solid, which cannot be moulded at all. At the end of the 28-day curing under water, the said mass yields almost its full target strength.

Be it the ordinary **concrete** (fine aggregate + coarse aggregate + cement), or sand-cement **mortar** (fine aggregate + cement), or just the **cement paste**, in which cement is the essential component responsible for inviting strength as a result of reaction in the presence of water, let us use the general term **concrete** or **mix** to mean all of them. Good workability is achieved in concrete when worked with within its initial setting time, after which, hardship of casting is experienced as the mass undergoes progressive setting towards the final setting time, and, at or around that time, the mass fully sets, and is transformed to a hard solid body in which workability reduces to nil.

Let us identify the time gap t, counted from the moment a concrete mass is prepared by adding water to the mixture of the dry ingredients, up to the time when it is cast in the mould, by the term **time lag**, (or simply **time**). In the light of the parameter, time lag t, the various stages of concrete, its workability and the corresponding strength can be described. When $t = 0$, the concrete is just prepared, fresh and fully green, and usually shows the maximum strength designed for a target value A_o. Beyond $t = 0$ and up to $t = t_i$, the concrete is ideally suited for casting and is having good workability, and yields strength A_i, which usually do not deviate much from A_o.

But if the prepared mass is left exposed in the open, and is taken for casting beyond the time lag t_i, the hardship of placing the mass in the mould is experienced owing to the ongoing of setting in it. Such mass is called **partially set** mass the strength A_t of which is seen to reduce greatly from A_o . It is worth noticing that, the strength of the **instantly prepared** mass at any identified time lag t, will yield the same value A_o, and is different from that of the **partially set** mass A_t observed at the time lag t.

Up to a time lag slightly beyond $t = t_i$, the partially set mass of strength A_t, though shows reduced strength, can still be used, if small strength difference is not of serious concern, and the mouldability of the mass does not pose serious difficulty. But as the time lag t of the partially set mass advances more towards $t = t_f$, the mass hardens greatly, the workability of which reduces at a faster rate, and around the final setting time, it is of no use for practical purpose. The strength of such hardened concrete is designated with A_f. Further, in reference to the time lag t, concrete, be it fresh, partially set or a hardened one, can be called by the generalized term, **partially set** concrete, the actual condition of which can easily be identified with the proper parameter t, as 0, t_i, or t_f. The corresponding strengths are then A_o, A_i, and A_f, respectively. As for common use, the mix showing strength A_o (for t between 0 and t_i), be called **fresh** mix, and that showing A_t (for t beyond t_i, and up to t_f) be called by **old** mix.

The term **strength** used here not only is restricted to designate **compressive strength**, but is broadly used to designate the other important strengths like, **tensile strength, shear strength , bond strength , flexure strength, impact strength**, etc.

True that the partially set concrete which has suffered a long period of exposure in the open before actual casting in the mould, should not be utilized for strength purpose in practice, and is to be discarded, but still it contains a good portion of the active part of the reactive cement in that mass, the strength of which could have been utilized with prudent judgement, if proper knowledge about the behaviour and strength characteristics of such partially set concrete is known. Such partially set concrete, particularly at time lag t, not far beyond the initial setting time, can be wisely and gainfully **reused** on blending it with another mix, be it a fresh one or even a partially set one, and which may be of the same or different mix type.

Selfing is a term attributed to the blending of two mixes, one being relatively **old** and the other being relatively **fresh**, of known strengths, each at their corresponding **time lags**, in certain **proportion**. **If the two mixes in blending are of the same mix type, they are in Selfing,** otherwise, **they are in Crossing.**

A **selfed** mass actually is a resulting mass ($\mathbf{Q = Q_o + Q_f}$), of two quantities of partially set concretes, namely, the **old mix** of quantity $\mathbf{Q_o}$, and the **fresh mix** of quantity $\mathbf{Q_f}$, blend in the ratio $\mathbf{r = Q_o / Q_f}$, by weight, and the condition of each constituent mix, again, is defined by its corresponding time lag parameter **t**. Interestingly, the same **selfed** mass **Q**, can very well be used to represent **any single** mix and its **condition**, which is controlled by the parameter **r**.

Thus, for $\mathbf{r = 0}$, the entire mix **Q** is a **single fresh** mix, in which the old part is totally absent ($\mathbf{Q_o = 0}$). Again, for **r = infinity,** the entire mix **Q** is a **single old** mix, in which the fresh part is totally absent ($\mathbf{Q_f = 0}$). Needless to say that, the **actual composite mix** comprising of two different mixes, ($\mathbf{Q = Q_o + Q_f}$), of course, is described with **r** assigned by any finite value lying between **0** and **infinity.** Further, the **condition** of both the single mixes (namely, **freshly prepared**, or, **partially set**, or, **hardened**), is described by the parameter **t**, as (**t = 0**), (**t =** t_i **to** t_f), or (**t = or >** t_f), respectively.

The description of any concrete (say, mix Type A), be it a single mass or a composite mass of two mixes, is made possible by adopting the generalized name as **Selfed concrete**, (or simply, **concrete**), and the **strength** of which by the proposed **two-parameter representation** through (**r** and **t**) as, **A(r, t)**, or, simply A_{rt}. Since, in blending, two quantities of mixes are involved, and each quantity of the mix has its own strength, the quantity (by weight) of the mix producing strength $A(r, t)$ or A_{rt} , is written within parenthesis as $(A(r, t))$, or, (A_{rt}). This is the essence of the **two-parameter representation** of **concrete**.

The terms **Selfing** and **Crossing**, as has been coined here, are not new, but are available in the standard text books of Botany, Zoology and Anthropology, used in an altogether different context. The pioneering works of Gregor John Mendel (1822 – 1884) which drew attention of the author as early as 1965, was on the study of the nature and character of the hybrid offsprings, resulting from the artificially crossed tall and dwarf varieties of garden peas, taken from two different progenies selected at two different filial generations. **Striking resemblance between the nature and behavior of such selfed or crossed offsprings, on one hand, and the strength and characteristics of the resulting partially set blended concretes, on the other side, inspired the author to search whether there exists any correlative link between these two phenomena.** Though Mendel's work is applicable to living objects (like plants), such expected link could not be established yet on the line of similar reasoning for concretes, which, after all, is a non-living object.

The main subject matter of the present Book is devoted to the various aspects of Selfing and Crossing, and is systematically arranged in four Chapters., *viz.*,

Chapter 1: ANY CONCRETE VIEWED AS PARTIALLY SET CONCRETE

Chapter 2: SELFING AND CROSSING CONCEPT

Chapter 3: SELFING EQUATION, ITS REPRESENTATIONS AND APPLICATION

Chapter 4: ANALYTICAL SOLUTION OF SELFING AND CROSSING PROBLEMS.

Some relevant discussion, typical observation on the subject presented in this Book, including critical comments and suggestion, if any of worth mentioning, is provided in CONCLUDING REMARKS, that follows the main four Chapters. Towards the end, a mathematical Theory, namely, **Spring Function** along with its Application and Calculation of actual problems, has been proposed for the first time in APPENDIX, which is expected to provide a suitable analytical tool for the future work to carry on, when more than two blended concrete mixes are in Selfing or Crossing.

The various aspects namely, the development of the concept, theoretical development and derivation of the mathematical formulation for the assessment of strengths, mathematical and mechanical and other analogous models of partially set concretes in Selfing or Crossing, illustration of the solution techniques with a large number of actual field problems with full length calculation are provided through these four Chapters. All topics in various Chapters are discussed with critical remarks in details, for better understanding of the unorthodox concept proposed for the first time by the author in the field of structural concrete. The predicted results using the theoretical formulae based on the proposed concept of Selfing and Crossing, can be seen to agree very closely with the corresponding experimental values. To the knowledge of the author, even today, there is no systematic study available on the partially set concretes, with particular emphasis on their reuse and assessment of strength.

The author carried out extensive research himself and also supervised a very large number of *B* Tech, *M* Tech as well as PhD theses on various aspects of the Selfing and Crossing, at the prestigious Indian Institute of Technology – Bombay, for the last thirty plus years. The constructive criticism and suggestion received during innumerable discussion with his students of different academic levels (B Tech, M Tech and PhD), and those during examination of a very large number of seminars and final defence, on this topic, both from his fellow faculty colleagues and external examiners, were kept in mind, as positive gain. Also experience gained over the years from the actual experimentation and screening of the enormous data collected from these, and a large number of field studies, enriched the knowledge of the author, which helped the author to formulate, develop and verify the technical theory on the strength and behaviour of partially set concretes. A small portion of the voluminous data obtained from the experiments has been utilized for the sake of illustration, wherever found suitable, in this present Book.

In this process, some terminologies, never existed before, in the field of Concrete Technology like, **Selfing**, **Crossing**, **Spring model for Partially set concretes**, **Negative spring**, (in which, a pull causes shortening, and vice versa), **Cumulative curing,** (such as, **Improper curing**, W_xA_y, first *x* days of curing in water, followed by air curing of y days, or, **Delayed curing**, A_xW_y, first *x* days of curing in air, followed by water curing of *y* days, such that, $x + y = c$ = total designed curing period), **Intermitant curing** (in which the continuity of continuous curing is replaced by stopping it at regular intervals, over the total curing period *c*), *etc*., have been brought to light and coined by the author.

Tests carried out from time to time and observation carried out in support of Selfing and Crossing are:

Concrete: of all varieties *viz*., **Cement paste** (mouldable, and non mouldable, *e.g.*, slurry), **sand-cement mortars**, ordinary **concretes** of all grades, and **high strength**

concretes. **Concretes with and without superplaticisers** (wherever applicable). **Fibre reinforced concretes** (with varying aspect rario and dosage of fibres), **Recycled aggregate concretes** (fine aggregate + recycled coarse aggregate + cement).

Curing condition (normal): Total curing under water for *c*-days as 7, 14 and 28 days.

Curing condition (Cumulative), *i.e.*, **Improper curing**, and **Delayed curing**

Strength: **Compressive** (Cube and cylinder), **tensile** (split cylinder, and diamond split using one half of the tested flexure specimen), **shear**, **flexure**, **bond** (concrete to concrete, and concrete to steel), and **impact strengths**.

Interfacial strength : **Interfacial bond** and **Interfacial tensile strength**

Tests were conducted for studying, verifying and establishing the analogous behaviour of Selfing and Crossing for studying bond strength, and tensile strength developed at the common interface of two layers in beams where, casting is done in two layers, in which the bottom layer is cast first, followed by the second top layer at a certain time lag, by varying mix grades. Cases considered: **concrete-on-concrete**, **concrete-net-concrete**, and **concrete-epoxy-concrete.**

Temperature cycle: Consists of heating the specimens and raised to a decided maximum temperature [30, 60 and 90 deg centigrade] in 2 hr, retaining there for 6 hr, then letting cool for next 16 hr, simulating day and night temperature cycle, for few thermal cycles, for prolonged time period.

A short list of the relevant publications in the form of Papers in reputed journals, Books, Theses (PhD, MTech and BTech), have been provided in SELECTED REFERENCE at the end. Since the behaviour and properties of concretes are available in any standard book of concrete technology, the reference of the text on this subject is omitted in this Book.

In this Book, the author has presented the two-parameter representation of partially set concretes and the mathematical formulation for the assessment of the blended concretes. The strength calculation of slurry mixed concrete, as often done in practice, and for which there exists no proper assessment method for strength uptil now, can now be easily done with the help of the Crossing formula. The mechanical spring model, and also other equivalent analogous models for depicting the behaviour of partially set concretes using the Selfing formula, derived as a closed form solution of the corresponding differential equation, is claimed to be the first of its kind, and is forwarded as a contribution in the field of concrete technology.

In the opinion of the author, the concept and subject matter presented in this book has opened new vista in the field of concrete technology, and is expected to be a helpful tool for the practicing engineers, research workers, and the students.

Nisith K. Bairagi

List of Figures

List of Figures

Notation

A = mix type
A_o = strength of OHC of the mix type A at $t = 0$
A_b = strength of blended composite mix = $((A_o) / (A_f))r$
A_i = strength of IPC of the mix type A, at $t = t_i$
A_t = strength of THC of the mix type A, at time lag $t = t\text{-}hr$ strength
A_f = strength of THC of the mix type A, at $t = t_f$
(A_t) = [strength within brackets] = quantity of mix showing strength A_t
$A_{\infty, t}$ = strength of THC for $r = \infty$, at time lag t
$A_{r, t}$ = strength of composite THC with blend ratio r, at time lag t
$A_{r, p}$ = strength at time lags $t = t_p$
$A_{r, q}$ = strength at time lags $t = t_q$
$A_{o, t}$ = strength of IPC for $r = 0$, at time lag t
$A_{r, t}$ = $((A_{\infty, t}) / (A_{o, t}))_r$
$\overline{A}_{rt}$ = complementary strength in Selfing = $A_{oo} - A_{rt}$
a_{rt} = reciprocal of $A_{r, t}$ $(= 1 / A_{rt})$
$\overline{a}_{ri}$ = vector representation of a_{ri}
$\overline{a}_{ri}$ = vector representation of a_{rf}
B = mix type
b = suffix
C = mix type, constant
CA = coarse aggregate
C_i = intercept at $t = t_i$
C_f = intercept at $t = t_f$,
COC = concrete-on-concrete
CNC = concrete-net-concrete
CEC = concrete-epoxy-concrete
D = constant
d_1, d_2 = distance
FA = fine aggregate
f = suffix indicating final setting time t_f, as in $A(r, f)$
IPC = instantaneously prepared concrete at any time t

i = suffix indicating initial setting time t_i, as in $A(r, i)$
K_1, K_2 = spring stiffness
k = suffix (denoting either i or f, indicating t_i or t_f)
M = concrete grade, (as in $M25$)
m, m_k, m_i, m_f = slope
n = integer
OHC = 0-hr concrete (= fresh concrete at $t = 0$)
p = suffix
q = suffix
Q = total quantity of blended mix, ($Q = Q_o + Q_f$)
Q_o = weight of THC (termed as **old** quantity)
Q_f = the weight of IPC (termed as **fresh** quantity)
q = suffix
R = blend ratio, function
R_1, R_2 = constant
R' = blend ratio
r = blend ratio = Q_o / Q_f
r_e = equivalent blend ratio
r_1 = blend ratio at $t = t_1$
r_2 = blend ratio at $t = t_2$
THC = t-hr concrete = time lag concrete
T, t = time, time lag
t_i = initial setting time
t_f = final setting time
T_i = time, time lag at t_i
T_f = time, time lag at t_f
w = deflection
w/c = water-cement ratio
X, Y = coordinates
x, y = coordinates
$\bar{x}, \bar{y}$ = non dimensional coordinates
α = angle
β = constant
θ = angle
λ = length / diameter ratio

Contents

UNIT 1
THERMODYNAMICS OR THERMAL SCIENCE

UNIT 1

Thermodynamics or Thermal Science

CHAPTER 1

Any Concrete Viewed as Blended Partially Set Concrete

1.1 INTRODUCTION

Concrete is prepared by thoroughly mixing the dry mixture of the ingredients viz., cement, sand (fine aggregate, *FA*), and coarse aggregate (*CA*) with the calculated amount of water. Cement is the only reactive material in the matrix responsible for developing strength, whereas, sand and coarse aggregates being inert materials, do not take part in chemical reaction and act only as solid filler material. If it is intended to term **concrete** as any mix containing cement as one of the ingredients, then cement paste, cement-sand mortar or concrete can be termed by the common name as concrete.

The matrix of wet green concrete is a heterogeneous mass of *CA*, *FA*, cement particles coated with water films, and air bubbles. Ideally the inter-particle spaces in between the bigger particles of the coarse aggregates should be filled up by the sand particles, and those spaces left over in between the fine aggregates should be occupied by the cement particles so as to get the mass a bonded single dense one. Proper vibration is aimed at achieving this goal in getting loose ingredients compacted together by driving out loose air pockets from the mass.

The development of strength in concrete depends on many factors like, water-cement (w/c) ratio, aggregate shape and size, duration of vibration, etc. Also it is true that denser is the mass more is the strength.

1.2 DELAYED CASTING, PARTIALLY SET CONCRETE

Concretes or cement-sand mortars or cement pastes develop strengths in the presence of water added to the dry mixture of the ingredients. The gain in the strength commences from the very instant of time when water is just added. This instant of time at $t = 0$ is identified as 0-hr, and at which the mix which is fresh and possess full workability, attains normally maximum strength. It is advisable that concrete thus prepared should be cast into mould without delay in order to get maximum strength out of it.

The time interval t between the preparation of the mix (at $t = 0$) and casting the same in the mould at elapsed time t (henceforth to be termed as **time lag** or for brevity, simply **time**) is a very influential parameter which plays a very major role from the point of development of strength in the mix. Such mix is to be termed as **time lag concrete** or **t-hr concrete** or **THC**, the strength as well as workability of which are bound to be affected adversely in comparison to that of the **fresh** or **OHC (O-hr concrete)**. Needless to mention that concretes instantly prepared at any instant of time t show the same strength as that of the 0-hr one, and they are identified by **instantly prepared concrete** or **IPC**. For a mix type A, thus the strengths of the OHC, IPC and THC can be written as A_o, A_o and A_t respectively, and those at $t = t_i$ (initial setting time) and $t = t_f$ (final setting time), are as A_i and A_f, respectively. In this sense, all concretes irrespective of being fresh OHC, IPC or actual THC can be broadly termed as **partially set concrete** or **THC**, which can be simply identified as the function of the parameter, namely, time lag t as

OHC as **partially set concrete THC, freshly prepared at t = 0**

IPC as **partially set concrete THC, instantly prepared at time t**

THC as **actual partially set concrete THC, prepared at time lag t**.

In actual concreting work, since some amount of time has to be provided right from the preparation of a mix to the casting in the mould, a time lag up to $t = t_i$ is usually tolerated, since both the strength as well as the workability do not differ appreciably from those of the 0-hr concrete. After this time limit, onset of setting in the mass commences inviting appreciable loss in the workability as well as reduction in strength. Concrete in the range of time lag beyond t_i cease to be green and fully plastic, since part of which has suffered a good deal of setting due to delay in casting, and so the THC in this range of time lag is identified as **partially set concrete**. As the time lag of casting approaches the final setting time t_f, the THC tends to become harder solid mass (primarily due to rapid evaporation of free surface water from the mass and accelerated setting process within the mix), rendering the mix unsuitable for casting (**Fig. 1.1**).

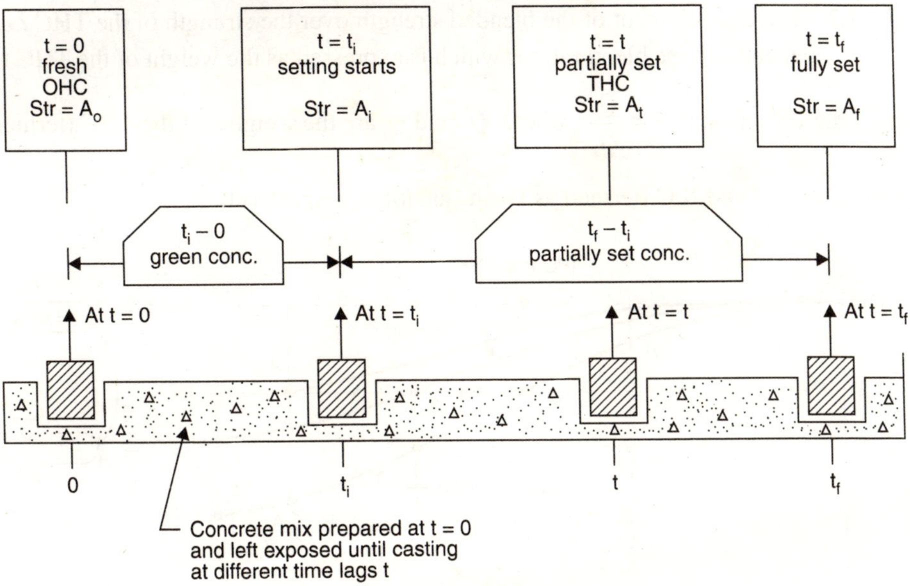

Fig. 1.1. Concretes cast at various stages of time lag

In field concreting works, it is often experienced that due to unforeseen circumstances like, displacement of formwork, break down of machineries, power failure, accidents, stoppage of work due to strikes etc., mixed concrete mass undergoes setting before being placed into the formwork. For time lag t between the interval 0 to t_i, the mixed mass may not invite any threat, and can be safely used. But for t far exceeding t_i and approaching t_f, the partially set THC being increasingly unserviceable, is not advised to be used, and is often declared as waste. The question arises as to what to do with huge amount of partially set mass of concrete–should it be discarded as a total waste, or is there any other way to gainfully reuse it?

1.3 FRESH, OLD AND BLENDED MIX

If the latter alternative is the choice, certain quantity of relatively fresher THC or freshly prepared IPC may be blended with the very partially set mass, and the composite mass thus produced may be reused. Such blending, though may not be a healthy practice, is often done in site concreting. Suppose that a certain mix type A shows the strength at time lag t (which is ahead of t_i) as **t-hr strength** = A_t. Obviously, this THC when blended with IPC at the same instant of time t (the strength of which is same as that of the **0-hr strength** = A_o), at a certain **weight ratio** r, the resulting **blended composite mix** would show a **strength** = A_b, which obviously would be greater than the t-hr strength A_t, and would lie between the two bounds viz., lower bound A_t and upper bound A_o. The ideal variation of the strength A_t vs. time lag t is drooping in nature

(**Fig. 1.2**). The enhancement of the blended strength over the strength of the THC *i.e.*, $(A_b - A_t)$ depends on the blend ratio r, which is expressed as the weight of the THC to that of the IPC. That is, $r = \frac{Q_o}{Q_f}$; where, Q_o and Q_f are the weights of the THC (termed as **old** quantity) and IPC (termed as **fresh** quantity), respectively.

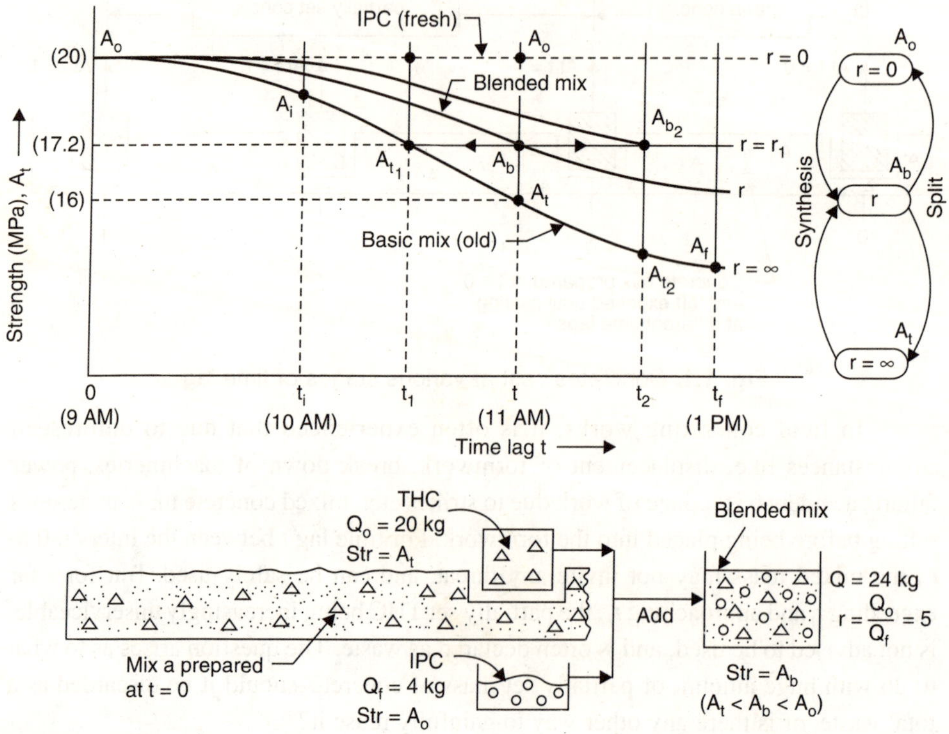

Fig. 1.2. Ideal strength-time variation of blended concretes

To observe how the blended strength A_b of a composite mix prepared out of the THC and IPC of some mix type A mixed in the weight ratio r (possessing strengths A_t and A_o respectively), is done at a certain time lag t, follow the following sequence:

Let the mix type A is M20 concrete so that the OHC strength at $t = 0$ is 20 Mpa, and the time lag of observation is $t = 2$ hr when the THC strength has reduced to 16 Mpa. The t_i and t_f values of the mix are 1 hr and 4 hr respectively. Let the blend ratio $r = 5$. At 9 AM some quantity of M20 is prepared and it is allowed to remain exposed to open atmosphere. This mix will constitute THC or **old** mix. Side by side, dry mixture of M20 is kept ready in another tray along with the requisite amount of water in a container. At 11 AM (after a lapse of 2 hr since 9 AM), prepare instantly fresh mix from the dry mixture on adding water to it. This constitutes IPC or **fresh** mix at 11 AM (*i.e*, at a time lag $t = 2$ hr). Now at this instant of time, take $Q_o = 20$ kg from the old mix (which has undergone 2 hr of exposure after it was prepared), and $Q_f = 4$ kg from the fresh mix, and thoroughly blend these two mixes together to get a total of $Q = Q_o + Q_f = 24$ kg mass, the strength of which is the blended strength A_b, which would show a strength of (say) 17.2 Mpa. This strength will be a little more than 16 Mpa but far less than 20 Mpa.

Since the relative proportion by weight of the two mixes viz., Q_o and Q_f in blending, possessing strengths A_t and A_o respectively, decides the increment of the strength $(A_b - A_t)$ over A_t, the corresponding quantities (*i.e.*, weights), are represented by writing the strengths within brackets for easy identification. That is,

$(A_o) = Q_o = 20$ kg [meaning 20 kg of the old mix A possessing strength A_o],

$(A_f) = Q_f = 4$ kg [meaning 4 kg of the fresh mix A possessing strength A_f],

so that $r = Q_o/Q_f = (A_o)/(A_f) = 5.$

Thus at any time lag, the total process of blending involving quantities, weight ratio and the corresponding strengths at one place can now be written symbolically as

$$A_b = ((A_o)/(A_f))_r$$

where, the **slash** symbol **will not be read as 'divided by', but as 'old quantity blended with fresh quantity in blend ratio'** r, with older and fresher quantities on the left and right sides of the slash sign, respectively.

It is important to note that, if $r = 0$, (for which $Q_o = 0$), the mix is understood to be either the single mass OHC or IPC, since it does not contain any old quantity at all.

On the other hand, if $r = \infty$, (for which $Q_f = 0$), the mix again represents the single mass of old or THC.

The actual blended mass of two mixes is described by any value of r between these two extremities, (*i.e*, $0 < r < \infty$).

In this sense, **whether a mix is actually blended composite or not, all single or blended masses can be described through the single parameter viz., blend ratio r.**

1.4 EQUIVALENT STRENGTH AND EQUIVALENT TIME

In **Fig. 1.2**, Curve 1 is the strength variation of THC for the basic Mix A alone in which no fresh quantity is blended at any time lag, so that the blend ratio $r = \infty$. Curve 2 which is a line parallel to the time lag axis represents the strengths of different IPC-*s* observed at the corresponding time lags having the same value A_o, and for which $r = 0$ (since the IPC-*s* are freshly prepared single mixes in which no partially set mixes are blended at all at any time). Curve 3 is the strength variation of actually blended composite mass of the THC and IPC in a blend ratio r whose value would lie between 0 to ∞.

With reference to **Fig. 1.2** it can be stated that the strength of the blended mix (with blend ratio r at time lag t), can be said to be **equivalent** to the strength of the basic mix A_{t1} (with blend ratio $r = \infty$ at time lag $t1$). This shows that **more is the elapsed time lag, more is the requirement of the fresh mix to be blended with the said partially set mix in order to retain the same higher strength as before**. Here since A_{t1} (for $r = \infty$) is the **equivalent strength** of A_b (for r), $t1$ is the **equivalent time** of t. Thus, $t1 < t$, if $A_b > A_t$.

Thus a very large quantity of IPC is required for blending with the THC to reach the enhanced strength of the composite mass to that of the fresh concrete, A_o. Practically it is not feasible, since the blend ratio r will approach to infinitely large value then. In other words if t increases, A_t rapidly reduces, and to get more and more additional strength in excess of A_t, the requirement of r becomes increasingly higher, demanding more and more fresh IPC for blending.

1.5 SYNTHESIS AND SPLIT

The new forms of symbolic representation of blending immediately leads to two types of operations, *viz*., **synthesis** and **split**, when read forward or reverse, respectively.

Synthesis: When two mixes of strengths A_t (for $r = \infty$) and A_o (for $r = 0$) are blended together in blend ratio r, the resultant single mix produces strength A_b, which is read on the strength curve for r.

Split: Any single strength A_b (observed on any strength curve for r) can be taken hypothetically as if composed of two mixes of strengths A_t (for $r = \infty$) and A_o (for $r = 0$).

The interesting feature of synthesis or split is that, at any time lag t ordinate, on the strength-time lag curves, any strength on the Curve 3 (possessing ratio r) can be read only in terms of those for Curve 2 ($r = 0$) and Curve 1 ($r = \infty$). In other words, **a single composite mix can be broken down to two mixes, IPC and THC, and vice versa, in which the corresponding strengths on r = 0 and r = ∞, are only involved**.

Referring to **Fig. 1.2**, the use of synthesis and split, and equivalent strength and equivalent time, can be demonstrated as follows.

Read **synthesis** as: x kg of the (single) mix A at time lag t when blended with y kg of the freshly prepared (single) mix A at the same time lag t, such that $x/y = r$, the resultant mix (composite) is to show strength A_b which is given by

$$A_b = ((A_t)/(A_o))_r. \tag{1.1}$$

Read **equivalent strength** and **equivalent time** as: A_b (read in r at time lag t) is equivalent to A_{t_1} (read in $r = \infty$ at time lag t_1).

That is, $$(A_b)_{\text{at } t} = (A_{t_1})_{\text{at } t = t_1}$$

Read **split** for A_{t_1} as: It is the strength of the composite formed out of p kg of the (single) basic mix at time lag t_1 in which no freshly prepared mix [*i.e.*, $q = (A_o) = 0$], is blended at all, so that $r = (A_{t_1})/(A_o) = p/q = \infty$. This justifies A_{t_1} on $r = \infty$ curve at $t = t_1$.

Similarly, since $A_{t_1} = A_{b_2} = A_b$, the corresponding split forms can be written as

$$[((A_{t_1})/(A_o)) \text{ for } r = \infty] = [((A_{t_2})/(A_o)) \text{ for } r = r1] = [((A_t)/(A_o)) \text{ for } r] \qquad (1.2)$$

Technically, the engineer-in-charge at site has to take hard decision whether to use such partially set concrete or not at all. If to allow such partially set concrete for use, how much reduction in strength could be tolerated for reuse on blending some fresher mixes (of the same or different mix type) to it? If so, up to what time lag and up to what blend ratio such blending could be allowed? Effectively, such decision calls for precise knowledge and calculation method of predicting the strength of two partially set concrete mixes blended in a certain weight ratio.

1.6 PRACTICAL PROBLEMS

In this book an attempt has been made to provide practical solution to a class of problems for which no calculation method exists even to day, on proposing altogether a new concept namely, **Selfing and Crossing** devised by the author. Some typical practical problems often encountered at construction site concerning partially set concretes in blending or those of similar kind have been cited below. These problems can be broadly grouped into two categories, as follows:

Category 1: Direct application of the concept in which two mixes of the same or of different mix types are blended together into a resultant single composite one, at a time lag t.

Category 2: Indirect or Analogous application of the concept in which the given problem can be transformed into an equivalent one to that of the Category 1, where physically mixing of two different quantities into a single one, may not be involved at all.

PROBLEMS OF CATEGORY 1

Example 1.1. ***Plastering work***

For a plastering work on a wall with cement-sand mortar, consider that some quantity of the mix is prepared at $t = 0$ hr, and the plastering work commences. After a considerable elapse of time t, which is well beyond the initial setting time of the mix, the quantity of the mortar mix is fully consumed. While the plastering work is in progress, the spill over mortar is getting collected at the bottom of the wall on plastic sheets or gunny bags. This spill over mass has become partially set in the mean time. To continue the remaining portion of the plastering work on the wall, fresh quantity of the said type of the mortar mix once again is prepared.

At this stage, if instead of using the freshly prepared mortar solely, the collected spill over mass is added to it, and then the blended mass of fresh and partially set mortars again is used for the said plastering work, the expected strength of the mortar mix would certainly bound to deviate from the designed strength of the fresh mortar.

Technically, one wants to know **what would be the final strength of the blended mortar, if the strength of both of the THC and IPC at time lag t, and the blend ratio r, of these two constituent mixes are given.**

Example 1.2. ***Large Mass Concreting work***

In a large concreting work say, in a dam construction, consider that a good amount of concrete is prepared at a place slightly away from the casting location, and is meant to be carried to the casting spot using various means like, truck, trolley, ropeway or even manually. If now the casting of the concrete is hampered due to some unforeseen circumstances such as, mechanical failure in machineries, disruption in electrical power supply, labour strike, accidents etc., and the delay persists for a considerable period of time exceeding the initial setting time of the mix, the huge mass becomes increasingly harder and harder. After that, if the situation returns to normal and the work is to resume, the concrete which has become partially set mass in that time gap will be of no use for casting, not only from workability but also from strength point of view too, and is bound to be discarded as unfit for use.

The question is what to do with such a huge partially set mass? Should it be thrown out as a stone hard solid waste, or could it be reused? If to throw out, another dumping ground is to be made available. Needless to say that, clearing and cleaning of the site, shifting and carrying the unused hard mass amount to additional cost and time. If instead, the intension is to gainfully reuse the partially set mass, then some fresh quantity of the same or slightly richer mix type can well be prepared and blended with it for the work to continue. However, in this process, the strength cannot be fully elevated to the designed target one, but at least to a comfortable margin. At times this may be acceptable to those types of work where dead weight of the structure is of prime concern, and the factor of safety in design takes care of the strength level thus achieved.

Technically, one wants to know **how much quantity of the IPC is required to blend with that of the THC at that time lag t, in terms of the ratio r, to get a predetermined target strength.**

Example 1.3. ***Casting work with slurry mixed concrete***

In a construction work, consider a situation where concrete could not be prepared at the very construction site, which may arise due to non availability of the materials in the near vicinity or clear ground space for installing concrete mixer, other machineries and storing the requisite ingredients. To overcome this situation, the obvious solution is that the concrete mix may be prepared at a distant place and is to be transported to the casting spot. Often such ready mix concrete is carried in open trucks or containers. Any unforeseen delay during transit which may result from events like, traffic jam, breakdown of the vehicle etc., initiates evaporation of the surface moisture from the mass as well as setting process. As a result, the mass starts loosing strength and workability, finally causing hardship in casting.

As often the crude practice goes, to regain workability, some amount of water is sprinkled on the mass prior to casting. However, in this process the effective water cement ratio is increased whose effect is to bring down the target designed strength. Again in order to compensate this reduction in strength, some quantity of dry cement is additionally mixed with the said mix. In effect, the treatment of the transported concrete with water and cement, one followed by the other, can be taken as equivalent to the treatment with **cement slurry** (or simply **slurry**), which is nothing but cement mixed in water.

Technically, one wants to know **what would be the final strength of the blended slurry mixed concrete if the strength of both of the THC (transported concrete) and IPC (cement slurry) at time lag t, and the blend ratio r of these two constituent mixes are given.**

PROBLEMS OF CATEGORY 2

Example 1.4. ***Casting of beams in layers***

For getting uniform characteristics and homogeneity throughout the mass of any structural concrete, it is essential that the entire body should be concreted at a stretch and without any break. Casting of concrete beams is no exception. Consider that in a project for casting of a number of beams, the concreting work starts at 10 AM ($t = 0$ hr).

After some portion of the beam moulds is filled with concrete from the bottom, the casting operation is disrupted all on a sudden at 10.30 AM ($t = 0.5$ hr) due to unforeseen circumstances (similar to those as described in Example 2), and the work comes to a halt. Let the concreting work is resumed, say at 12.30 PM ($t = 2.5$ hr), and by that time, onset of setting has already commenced in the concrete mass inside the beam moulds. Though not permitted, let the casting work of the remaining portion of the beams starts with freshly prepared concrete mix (of the same or of different mix type) over the partially set first layer, and the work completes at 1.30 PM.

Thus each beam is actually cast in two layers. At the said instant of time lag t, the top layer (which is the IPC) rests over the bottom one (which is the THC at time lag $t = 2.5$ hr). The physical as well as the strength properties of the concrete of any such individual layer is bound to differ appreciably from the corresponding properties of the other. Bond plays an important role in holding together two such partially set concrete layers in order to behave as an integrated mass. The monolithicity of such individual concrete layers of partially set concretes at a certain time lag depends greatly on the bond developed at the common interface, which of course, deviates from the corresponding bond at the same level had the beam is cast at one stretch from the bottom to the top, say with OHC only.

The interfacial strength in question, is the bond developed at the common surface of contact between the two concrete layers cast at a certain time lag. This bond is identified as **concrete-on concrete** bond, or simply **COC bond**, and is different from the bond used in reinforced concrete design, which actually is the **metal-matrix** bond. **The COC bond strength may be the shear strength or the tensile strength developed at the interface of two time lag concretes in contact, and will be termed as A_I.**

Technically, one wants to **assess A_I in terms of the bond strengths of the constituent individual layer of the bottom THC, A_t, and that of the top IPC, A_o, in relation to the time lag t, for different variations of mix grade M.**

Example 1.4 can be viewed as analogous to Example1.1 or Example 1.2.

Example 1.5. ***Fibre reinforced Concrete***

Concrete when mixed with fibers in suitable proportion of the cement concrete matrix, is often found to be very advantageous in situations where enhancement of tensile, shear and impact resistance as well as delay of the onset of cracks leading to untimely deterioration or early collapse are concerned. Depending on the source or origin wherefrom they are derived, fibers may be termed as natural fibers, chemical fibers, metallic fibers etc.

In reality, a structure will have to withstand continuous cycles of heating during the day followed by cooling during the night, repeatedly occurring throughout its lifetime. Consider the concreting work of a structure to stand in the open for which it is decided that fiber reinforced concrete be gainfully used instead of ordinary concrete, and for which the proper choice falls on steel fibers. The concreting proceeds as usual up to a certain time, and then unforeseen situation as described as in Example 1.2 appears.

1.7 TWO-PARAMETER REPRESENTATION OF STRENGTH

In reference to the cited Examples (Category 1), it can be observed that

(*i*) **There are two mixes in blending.**

(*ii*) **The two mixes may be of same mix type or of different mix types.**

(*iii*) **One of the mix is older than the other** (viz., **old** and **fresh**), **and are of known strengths.**

(*iv*) **The older mix is at time lag t, the other one is at time lag t = 0, so that the strengths are A_t and A_0 (for the same mix type A), or, A_t and B_0 (for different mix types A and B).**

(*v*) **The proportion of the old to the fresh mix is in the blend ratio r.**

(*vi*) **The resultant blended mix is a single composite mass whose strength is to be expressed in terms of the blend ratio r, and the time lag t.**

Thus, representing at any instant of time lag t, a quantity of partially set mix by (A_t) of strength A_t, when blended with a quantity of instantly prepared mix (A_o) of strength A_o, at the same instant of time lag t, in a blend ratio r, such that, $r = (A_t)/(A_o)$, the strength of the blended composite mass can be expressed as $A(r, t)$ or $A_{r,t}$ or simply, A_{rt}. This is precisely **two-parameter representation of strength.**

Admitting this new form of representation of strength (*viz.*, $\mathbf{A_{rt}}$ – **representation**), any **mix can now be brought under broader definition by recognizing it as a partially set composite mix on suitably adjusting the values of r and t,** as follows:

Any mix, be it single or composite, can be termed as a blended composite mix defined by the blend ratio r, so that r = wt of old / wt of fresh = Q_0/Q_f, such that

if $r = 0$,	the composite mix is an **instantaneously prepared fresh** mix at any time lag t,
if $r = \infty$,	the composite mix is a partially set **basic** single mix at any time lag t,
if $0 < r < \infty$,	the composite mix is **actually a blended one**, composed of the **old** part and **fresh** part at any time lag t.

With this definition of mix, **any mix, be it fresh, freshly prepared or partially set one, can be termed as partially set mix, defined by the time lag t,** so that for $t = 0$, the mix actually is a **fresh** one and **workable,**

for $0 < t < t_f$, the mix actually is a **partially** set one and **partially workable,**

for $t \leq t_i$, the mix is **workable**

for $t = t_f$, the mix is **fully set**.

Thus, referring to **Fig. 1.2**, we can rewrite the strengths of THC *i.e.*, A_t of the basic mix (for $r = \infty$), blended mix A_b (for r), and of IPC *i.e.*, A_o (for $r = 0$), at any time lag t as

$A_{\infty, t}$ = strength of THC for $r = \infty$

$A_{r, t}$ = strength of THC for r

$A_{o, t}$ = strength of IPC for $r = 0$,

and, the with the help of this notation, the **synthesis** and **split** processes (Art 1.5), can be rewritten as

Synthesis: blending of two quantities ($A_{\infty, t}$) and ($A_{o, t}$) in the ratio r, to form a single mass of strength A_{rt}, that is

$$((A_{\infty, t})/(A_{o, t}))_r = A_{r, t} \tag{1.3}$$

Split: a single mass of strength $A_{r, t}$ is considered as if formed of blending two quantities

($A_{\infty, t}$) and ($A_{o, t}$), in the ratio r, that is

$$A_{r, t} = ((A_{\infty, t})/(A_{o, t}))_r \tag{1.4}$$

The **two-parameter representation of strength** and the use of tools like, **synthesis** and **split** form the very basis of the **Selfing** and its generalized version viz., **Crossing** concept, which is the subject matter dealt elaborately in this book through the following chapters.

CHAPTER 2

Selfing and Crossing Concept

2.1 ABOUT SELFING AND CROSSING

Selfing is a term attributed to the blending of two different individual mixes of the same mix type but of different r and t values into a single composite mass, which henceforth be called as the **selfed mass**, and the corresponding strength of which be termed as **selfed strength.**

Crossing, on the other hand, is the generalized version of selfing, where the two mixes in blending are of different types, and the corresponding terms are **crossed mass** and **crossed strength.**

Example 2.1. *(in Selfing)*

20 kg of 1 : 3 fresh mortar (*i.e.*, OHC, $Q_f = 20$ at $t = 0$) is blended with 40 kg of same 1 : 3 partially set mortar considered at time lag $t = 2.5$ hr (*i.e.*, THC, $Q_o = 40$ at $t = 2.5$), to form a blended mix of total mass = 60 kg.

Both the mixes, each of type A is 1 : 3 mortar

Strength of fresh mix = $A_{0,\,0} = A_{0,\,2.5}$ (since strength of IPC and OHC is same)

Weight of fresh quantity = $(A_{0,\,2.5}) = Q_o = 20$ kg

Strength of old mix = $A_{\infty,\,2.5}$

Weight of old quantity = $(A_{\infty,\,2.5}) = Q_o = 40$ kg

The two quantities are blended in ratio, $r = (A_{\infty,\,2.5})/(A_{0,\,2.5}) = 40/20 = 2$

Time at which the two mixes are blended, $t = 2.5$ hr

Hence using synthesis (1.4), the selfing process symbolically is written as

$$A_{2,2.5} = ((A_{\infty,\,2.5})/(A_{0,\,2.5}))_2 \quad (2.1)$$

(**Note:** Since the strength-time lag curves for the mix type A for various blend ratios are valid for any two A-type mixes in blending (so also is the case for the said two mixes here), the resultant selfed strength can be read as that for the mix A for which the proper (r, t) set is to be determined. If the time lag t is decided to be the same at $t = 2.5$, it requires then to locate the proper value of r. The synthesis concept provides that $r = 2$. That is, on the THC curve for $r = 2$, and at $t = 2.5$, the strength observed is the required selfed strength $A_{2,\,2.5}$.)

In **Fig. 2.1**, the selfing process and the selfed strength is described.

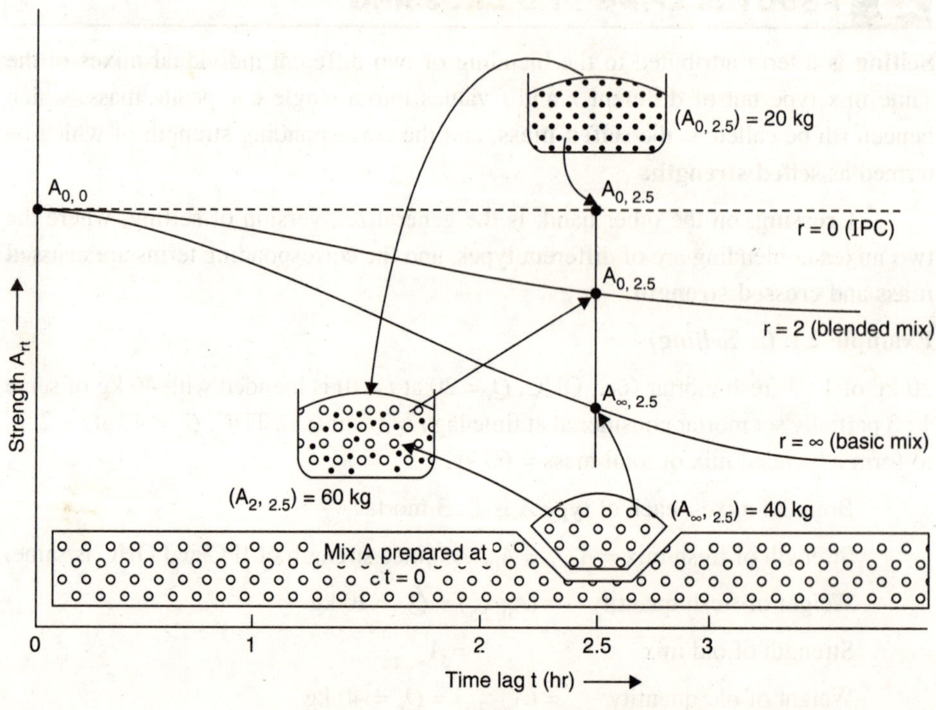

Fig. 2.1. Description of pure selfed strength

Example 2.2. *(in Crossing)*

80 kg of M30 partially set concrete considered at time lag $t = 3$ hr (*i.e.*, mix A is THC, $Q_o = 80$ at $t = 3$), is blended with 50 kg of M20 fresh concrete (*i.e.*, mix B is OHC, $Q_f = 50$ at $t = 0$), to form a blended mix of total mass = 130 kg.

Mix A is M30 concrete, and mix B is M20 concrete.

Strength of fresh mix $= B_{0,\,0} = B_{0,\,3}$

Weight of fresh quantity $= (B_{0,\,3}) = Q_o = 50$ kg

Strength of old mix $= A_{\infty, 3}$

Weight of old quantity $= (A_{\infty, 3}) = Q_o = 80$ kg

The two quantities are blended in ratio, $r = (A_{\infty, 3})/(B_{0, 3}) = 80/50 = 1.6$

Time at which the two mixes are blended, $t = 3$ hr

Hence, the crossing process is symbolically written as

$$C_{R, T} = ((A_{\infty, 3})/(B_{0, 3}))_{1.6} \quad (2.2)$$

where, the final crossed strength $C_{R, T}$ means that the strength, if observed in a third mix type C (say M25), it would be that corresponding to $r = R$ and $t = T$.

Note: However, in reality, the $C_{r, t}$ - curves must be available in hand. Instead, the resultant crossed strength may be read either in mix A or mix B, the strength-time lag curves of which are considered to be known beforehand. Suppose mix A is selected for the purpose, then it remains only to fix the time lag and getting the suitable blend ratio $R = R_1$.

If the time lag is fixed at $T = 3$, then the resultant crossed strength is given by $A(R_1, 3)$ or $A_{R_1, 3}$. Similarly, if the crossed strength is read in mix B, then for $T = 3$, it would be $B(R_2, 3)$ or $B_{R_2, 3}$, where $R = R_2$ is the suitable blend ratio.

Recollect that, the strength of a single mix in selfing concept can be expressed as the blending of two mixes of similar kind one part of which is old and the other part is fresh, at a time lag t as A_{rt} or $A_{r, t}$ or $A(r, t)$, where r can take either 0 or ∞, at any time lag t. That is

$A_{o, t}$ = strength of single IPC = A(IPC), at time lag t.

$= A_{o, o}$ = strength of single OHC (= strength of fresh

$= A$(OHC), at time lag $t = 0$.

$A_{\infty, t}$ = strength of single THC = A(IPC), at time lag t.

= strength of old = A(THC), at time lag t.

Thus, A_{r_1, t_1} = strength of actual composite mix of two mixes of strengths of the old part A_{∞, t_1} and that of the fresh part $A_{o, t1}$, blended in ratio $r = (A_{\infty, t_1})/(A_{o, t_1})$ at time lag t_1, where r is any value between 0 and ∞, but excluding these two limiting values.

Consider two such individual partially set mixes viz., mix A and mix B of strengths A_{r_1, t_1} and B_{r_2, t_2} respectively, which are blended in weight ratio $r = (A_{r_1, t_1})/(B_{r_2, t_2})$, at some suitable time. The synthesis process is expressed symbolically as $((A_{r_1, t_1})/(B_{r_2, t_2}))_r$. It represents crossing, since mix A and mix B are different. It will represent Selfing, if mix A and mix B are identical. The synthesis process is then $((A_{r_1, t_1})/(A_{r_2, t_2}))_r$.

2.2 SUBGROUPING OF SELFING AND CROSSING

With these, further subgrouping of selfing and crossing can be done as **General**, **Pure** and **Absolute**.

(*i*) **Absolute Case :** $r_1 = r_2$, and $t_1 = t_2$.

(*ii*) **Pure Case :** $r_1 \neq r_2 (= 0)$, and $t_1 = t_2$.

(*iii*) **General Case :** $r_1 \neq r_2$, and $t_1 \neq t_2$.

In **Fig. 2.2** the blending of two mixes *A* and *B* at different time lags is shown through line diagram, where *A*, *P* and *G* represent absolute, pure and general case, respectively. If mix *A* is identical to mix *B*, it is selfing, otherwise crossing.

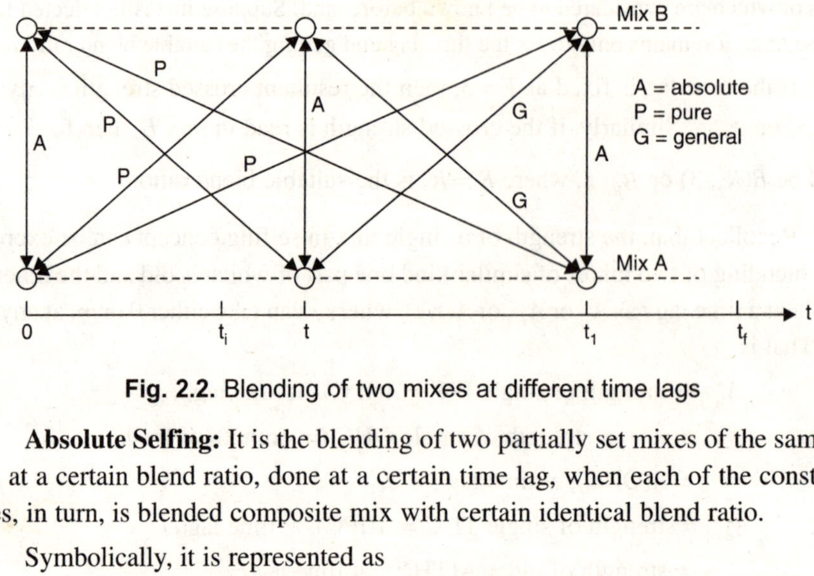

Fig. 2.2. Blending of two mixes at different time lags

Absolute Selfing: It is the blending of two partially set mixes of the same mix type, at a certain blend ratio, done at a certain time lag, when each of the constituent mixes, in turn, is blended composite mix with certain identical blend ratio.

Symbolically, it is represented as

$$((A_{r_1,t_1})/(A_{r_1,t_1}))_r \text{, at time lag } t.$$

Since both the constituent mixes in blending depict identical strengths A_{r_1,t_1}, which are independent of the quantities from which they are observed, and of the blend proportion they are mixed, the strength of the final selfed mass is the same as that of the constituent mixes. So the determination of strength through absolute selfing does not serve any purpose, and is of academic interest only.

Example 2.3.

80 kg of M30 partially set concrete considered at time lag $t = 3$ hr is blended with 50 kg of the same M30 partially set concrete considered at the same time lag $t = 3$ hr, to form a blended mix of total mass = 130 kg.

Here, in the first mix M30, $Q_O = 80$ kg, and in the second mix M30, $Q_f = 50$ kg, Total $Q = Q_o + Q_f = 130$ kg, which is also solely M30 concrete.

The strength of the mix whether taken from the first quantity of 80 kg, or from the second quantity of 50 kg, or from the final selfed composite quantity of 130 kg, has to be the same.

Absolute Crossing: Mix *A* and mix *B* are different but the corresponding blend ratio and time lag of each mix in blending are identical.

Example 2.4.

Consider Example 2.3 as it is, with the first mix M30, and the second mix M20, with other data remaining same.

Pure Selfing: It is the blending of two partially set mixes of the same mix type, at a certain blend ratio, done at certain time lag, in which one of the constituent mixes is really a blended composite mix with a certain blend ratio, done at a certain time lag, while the other one is the instantaneously prepared mix at that time lag.

Symbolically, it is represented as

$$((A_{r_1, t_1})/(A_{o, t_1}))_r \tag{2.3}$$

If the former mix happens to be the single basic mix (for which $r = \infty$), then the pure selfing operation is directly represented by

$$((A_{\infty, t_1})/(A_{o, t_1}))_r = A_{r, t_1} \tag{2.4}$$

which can be identified as the symbolic representation for synthesis (1.3).

Examples 1.1–1.5, and Example 2.1, are the illustrations of pure selfing. The latter example has been described in **Fig. 2.1**.

Pure Crossing: It is similar to pure selfing case but the two partially set mixes in blending are different. One of the mix *A* (old) is with (r_1, t_1), while the other mix *B* (IPC) is with $(0, t_2)$ which is synonymous with *B* (IPC) with $(0, t_1)$, and also with *B* (OHC) with (0, 0).

General Selfing: It is the blending of two partially set mixes of the same mix type, done at a certain blend ratio, and observed at a certain time lag, while, each of the constituent mixes, in turn, is a blended mass considered at their respective time lags.

Symbolically, it is represented as

$$((A_{r_1, t_1})/(A_{r_2, t_2}))_r, \text{ at any suitable time lag } t, (t_1 > t_2) \tag{2.5}$$

The split of the constituent mixes, which when read in pure selfing (2.4), are

$$A_{r_1, t_1} = ((A_{\infty, t_1})/(A_{o, t_1}))_{r_1}, \text{ at time lag } t_1, \text{ and} \tag{2.6}$$

$$A_{r_2, t_2} = ((A_{\infty, t_2})/(A_{o, t_2}))_{r_2}, \text{ at time lag } t_2.$$

Example 2.1 serves as the illustration for the constituent mixes.

The operational method of a typical general selfing scheme is described in **Fig. 2.3** as follows:

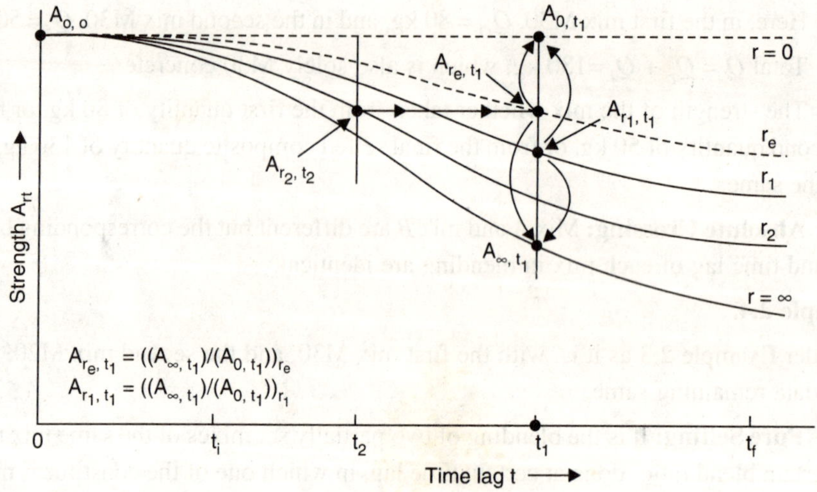

Fig. 2.3. General selfing scheme

Select that the final selfed strength will be observed at $t = t_1$.

The older strength is A_{r_1,t_1}. Using split in pure selfing,

$$A_{r_1,t_1} = ((A_{\infty,t_1})/(A_{o,t_1}))_{r_1}$$

The fresher strength is A_{r_2,t_2}, which when transferred on t_1, becomes A_{r_e,t_1}, where r_e is the corresponding blend ratio, (yet to be found out). That is, $A_{r_2,t_2} = A_{r_e,t_1}$.

Again using split in pure selfing, $A_{r_e,t_1} = ((A_{\infty,t_1})/(A_{o,t_1}))_{re}$

Thus the general selfing can be always written in terms of two pure selfing break-ups of two sets of strengths with (∞, t_1) and $(0, t_1)$ combinations on any instant of time lag suitably selected. The resultant selfed strength is then A_{R,t_1}. It requires to find the value of R from the given blend ratio values r_1, r_2 and r. Since r_2 (on t_2) is expressed in terms of its equivalent r_e (on t_1), R then can be written in terms of r_1, r_e and r.

The time lag, to be selected, may or may not be on the time lags of the individual mixes (viz., t_1 and t_2). It may be a third one any where between $t = 0$ to t_f, and may even be at a hypothetical one for $t > t_f$.

The guide lines for the proper selection of the time lag t at which the final selfed strength is intended to be read, is as follows. Refer to **Fig. 2.4** (with $t_1 > t_2$).

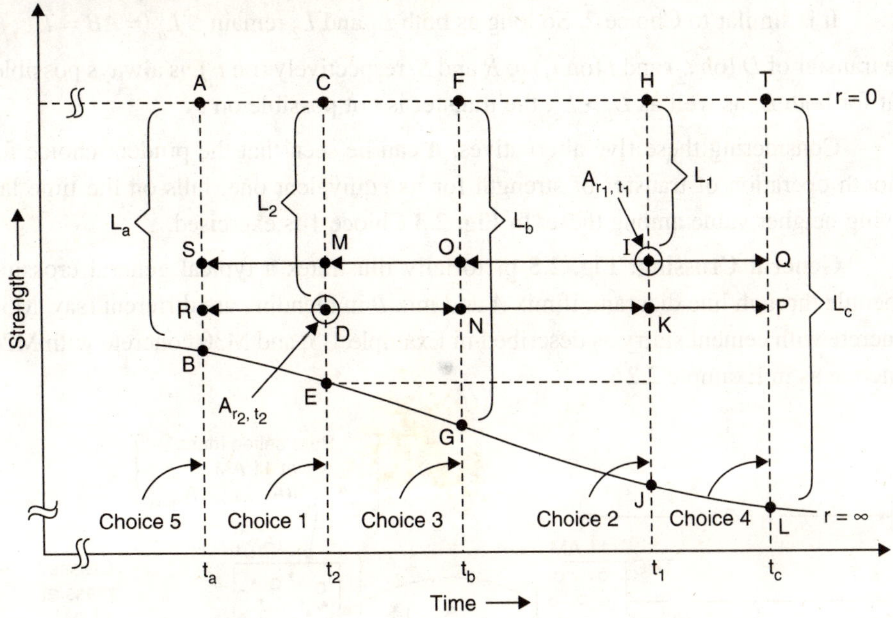

Fig. 2.4. Selection of time lag for reading selfed strength

Choice 1. t is at t_1:

A_{r_1,t_1} remains at I.

Project A_{r_2,t_2} from D to K. Since HJ (= L_{∞,t_1}) > CE (= L_{∞,t_2}), for any position of r_2;($0 < r_2 < \infty$) on t_2, the transfer of D (on t_2) to K (on t_1) is always possible.

Choice 2. t is at t_2:

A_{r_2,t_2} remains at D.

Project A_{r_1,t_1} from I to M. So long as L_1 (= HI) ≤ CE (= L_{∞,t_2}), irrespective of any position of r_1 (on t_1), the transfer of I (on t_1) to M (on t_2), is possible. But for L_1 > CE, the transfer of I to M is not possible.

Choice 3. t is in between t_1 and t_2:

$$HJ\ (= L_{\infty,t_1}) > FG\ (= L_{\infty,t}) > CE\ (= L_{\infty,t_2}).$$

The projection of A_{r_2,t_2} (from D) to N (on t_b) is always possible, since $FG > CE$.

But the projection of A_{r_1,t_1} (from I) to O (on t_b) is always possible for $L_1 \leq FG$, and not possible for $L_1 > FG$.

Choice 4. t is at t_c:

It is similar to Choice 1, since L_c (= L_{∞,t_c}) > HJ (= L_{∞,t_1}) > CE (= L_{∞,t_2}). The transfer of D (on t_2) and I (on t_1) to P and Q, respectively (on t_c), is always possible.

Choice 5. t is at t_a:

It is similar to Choice 2. So long as both L_1 and L_2 remain $\leq L_a$ (= $AB = L_{\infty, t_a}$), the transfer of D (on t_2) and I (on t_1) to R and S, respectively (on t_a), is always possible. But for both L_1 as well as $L_2 > L_a$, the transfer is not possible on t_a.

Considering these five alternatives, it can be seen that the prudent choice for smooth operation of transfer of strength for its equivalent one, falls on the time lag having heigher value among these. In **Fig. 2.3** Chioce 1 is exercised.

General Crossing: Fig. 2.5 pictorially illustrates a typical general crossing schedule through line diagram, if mix A and mix B in blending are different (say, M30 concrete with cement slurry as described in Example 1.3, and M30 concrete with M20 concrete as in Example 2.2).

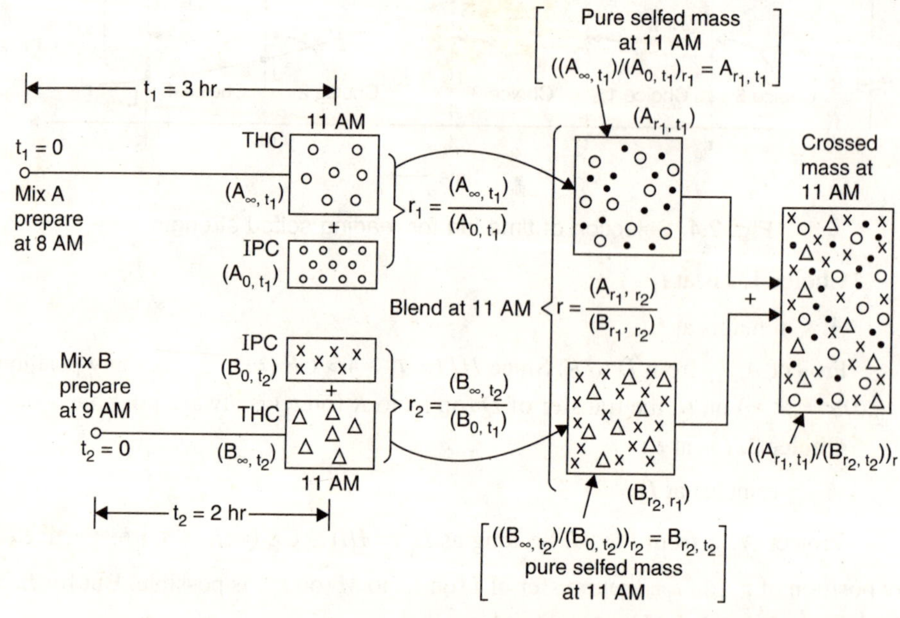

Fig. 2.5. General crossing schedule

If mix A and mix B are identical, the same **Fig. 2.5** will be read as the description for general selfing schedule.

That the general selfing is the particular form of general crossing is illustrated in **Fig. 2.6** for $((A_{r_1, t_1})/(A_{r_2, t_2}))_r$. The equivalent strength of one of the mix (say, mix B for $r = r_2$ and at $t = t_2$) is sought for in terms of the strength of the other (say, mix A for $r = r_e$ and at $t = t_1$). That is, $B_{r_2, t_2} = A_{r_e, t_1}$.

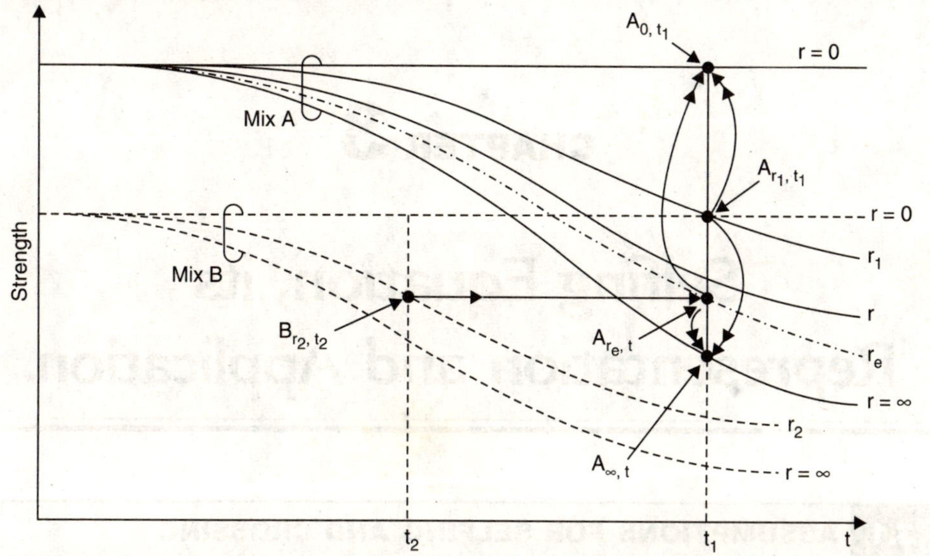

Fig. 2.6. General crossing viewed as general selfing scheme

Then the operation of general crossing simply becomes the problem of general selfing as $((A_{r_1,t_1})/(A_{r_e,t_1}))_r$, since both the strengths of the two mixes in blending are now of mix type A, and each one when split in pure selfing, can be described in terms of $A(\infty, t_1)$ of the basic THC, and $A(0, t_1)$ of the IPC at t_1, blended in ratios r_1 and r_e respectively.

Whether selfing or crossing, there are **six** quantities to handle viz.,

(*i*) **the strength of the two individual constituent mixes (old and fresh) and that of the blended one (*e.g.*, A_{r_1,t_1}, B_{r_2,t_2} and $C_{R,T}$)**

(*ii*) **the blend ratios (R) and (r)**

(*iii*) **the time lag (T).**

If any five of these are given, the remaining one can be found.

CHAPTER **3**

Selfing Equation, its Representation and Application

3.1 ASSUMPTIONS FOR SELFING AND CROSSING

The framework of assumptions on which the technical theory of selfing as well as crossing are based, may be stated as follows.

1. Any individual mix is considered as a partially set one identified with the time lag t.
2. Any individual mix is considered as a blended mass of two partially set mixes of the same mix type, one part of which is the fresh and the other part is the old, identified with the blend ratio r.
3. The strength of any individual mix is considered to be the function of two parameters viz., r and t, whereas, the influence of the other parameters like, compaction, curing type, temperature etc., are said to remain constant.
4. The strength of any mix attains its maximum value at zero time lag (*i.e.*, when it is OHC), from which, it gradually droops down to minimum at final setting time.

   Further, the strength of OHC (at $t = 0$) is synonymous with that of IPC (at any t), *i.e.*, $A_{o,o} = A_{o,t}$.
5. The shape of the individual strength-time curve for any variation of r is practically linear between the time lag bounds of t_i and t_f, so that the slope m at any location is same through out.
6. The strength of an actually blended mass depends primarily on the overall proportion of the old and fresh parts contributed from those of the corresponding parts of the two constituent partially set masses in blending.

[Discussions on these assumptions and suggested modifications for improvement, if any, are provided latter.]

3.2 SELFING EQUATION

In **Fig. 3.1** typical idealised variation of A_{rt}- curves for a certain mix A is shown. At any time lag t, the strength of the basic mix (corresponding to $r = \infty$) is $UQ = A_{\infty, t}$. The same basic mix when selfed with freshly prepared mix (whose strength is given by $US = A_{o, t}$) in blend ratio r, such that $r = (A_{\infty, t})/(A_{\infty, t})$, is given by $UP = A_{\infty, t}$.

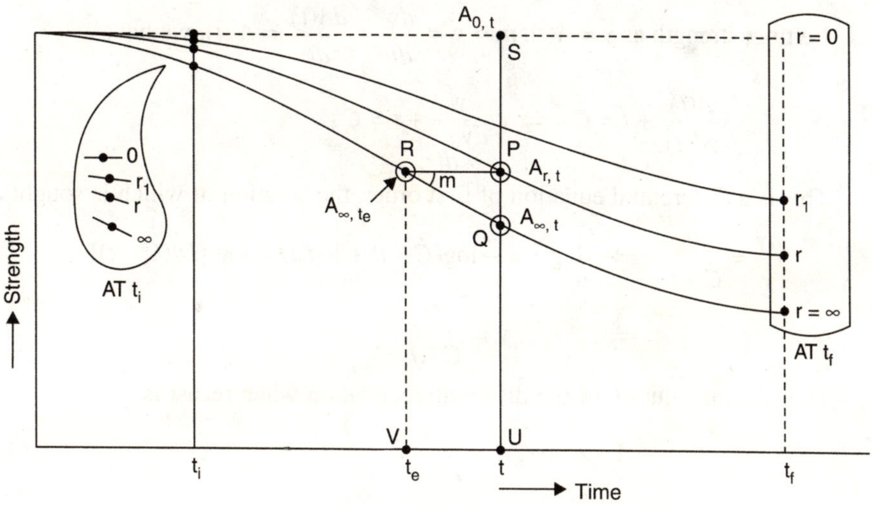

Fig. 3.1. Equivalent strength corresponding to equivalent time lag

The equivalent strength of $A_{r, t}$ at P (on r at t) = $A(\infty, t_e)$ at R (on $r = \infty$ at $t = t_e$)

The strength increment of the selfed mass over the strength of the basic partially set mass at t is $PQ = A(\infty, t_e) - A(\infty, t)$

Consider the triangle PQR between the time lag t and its equivalent t_e (with the assumption that RQ is linear),

$$m = \text{slope of } RQ = \frac{PQ}{PR} = \frac{A(r,t) - A(\infty, t)}{t - t_e} = \frac{A(\infty, t_e) - A(\infty, t)}{t - t_e}$$

$$= C = \text{constant}$$

where, $A(\infty, t_e)$ = equivalent strength of $A(r,t)$ observed at t_e.

$$t - t_e = \frac{A(\infty, t_e) - A(\infty, t)}{m} = C$$

$$\Rightarrow \qquad \frac{A(\infty, t_e)}{m} + t_e = \frac{A(\infty, t)}{m} + t = C$$

Note: The strength on any arbitrary curve r (at t) is projected here on curve $r = \infty$ (at t_e). So this statement, expressed for strengths in terms of the parameters on the curve $r = \infty$, relating t and its equivalent t_e, is valid for this curve (namely, for $r = \infty$). Generalising, it can be inferred that, this statement also is equally true on any r- curve.

Thus for any particular variation of r, we have

$$\frac{strength,\ A(t)}{slope,\ (m)} + time,\ (t) = const.\ (C) \tag{3.1}$$

Letting, strength as $y = A(t)$, $m = y' = \frac{dy}{dt} = \frac{dA(t)}{dt} = A'(t)$, we have

$$\frac{A(t)}{A'(t)} + t = C \quad \Rightarrow \quad \frac{y}{dy/dt} + t = C$$

This is a differential equation of first order, the solution of which is sought as

$$\frac{dy}{y} = \frac{dt}{C-t} \quad \Rightarrow \quad \log y = -\log(C-t) + \log D = \log\,[D/(C-t)]$$

$$\Rightarrow \qquad y = \frac{D}{C-t}$$

The general solution of the differential equation when recast is

$$A_{r,t} = A(r, t) = \frac{D}{C-t} \tag{3.2}$$

where, C and D are the two arbitrary constants to be evaluated with the help of given boundary conditions, viz.,

at $t = t_i$, $A(r, t) = A(r, t_i) = A(r, i)$,

and at $t = t_f$, $A(r, t) = A(r, t_f) = A(r, f)$.

(where, for brevity, **sometimes** instead of t_i and t_f, their suffixes i and f, respectively, have been used to represent the said time lag parameters.)

Thus solving,

$$C = \frac{A_{r,i}\,.\,t_i - A_{r,f}\,.\,t_f}{A_{r,i} - A_{r,f}}, \quad D = -\frac{A_{r,i}\,.\,A_{r,f}\,.\,(t_f - t_i)}{A_{r,i} - A_{r,f}} \tag{3.3}$$

With these constants the required selfing equation is

$$A_{r,t} = \frac{A_{r,i}\,.\,A_{r,f}\,.\,(t_f - t_i)}{t\,.\,(A_{r,i} - A_{r,f}) + A_{r,f}\,.\,t_f - A_{r,i}\,.\,t_i} \tag{3.4}$$

This equation is the **selfing equation** or **A_{rt} - equation**.

The **two alternative forms** of this selfing equation are

$$\frac{t_f - t_i}{A_{r,t}} = \frac{t - t_i}{A_{r,f}} + \frac{t_f - t}{A_{r,i}} \tag{3.5}$$

and, $$(t_f - t)\left[\frac{1}{A_{r,t}} - \frac{1}{A_{r,i}}\right] = (t - t_i)\left[\frac{1}{A_{r,f}} - \frac{1}{A_{r,t}}\right] \quad (3.6)$$

In the selfing equation [(3.4) or (3.5) or (3.6)], which are derived for a given mix A, and for any particular variation of r, $(0 \le r \le \infty)$,

given are (five quantities): setting times t_i and t_f, the corresponding strengths at these two instant of time lags viz., $A_{r,i}$ and $A_{r,f}$, respectively, and also the decided time lag t,

to find (one quantity): selfed strength $A_{r,t}$ at this time lag t.

Note: Since the selfing equation derived here, is only valid for any particular variation of the blend ratio r, it can readily calculate the strength $A_{r,t}$ at any intermediate time lag t, for that variation of r, if only two strengths viz., $A_{r,i}$ and $A_{r,f}$ at the two setting times viz., t_i and t_f respectively, are given.

Referring to the fact that the strength of any mix is defined with two parameters viz., (r, t), it is important to note that, the selfing equation (3.4), even though which is obtained as the closed form solution of the differential equation (3.1), cannot be directly used, as it is, for practical purpose. The reason is that, even if it contains the parameter t, it has no room for the other parameter r. However, this deficiency can easily be eliminated if each of the two strengths on the r-curve viz., $A_{r,i}$ and $A_{r,f}$ observed at t_i and t_f, respectively, are expressed as suitable function of the parameter (blend ratio) r. This can be best done taking guide lines from the actual plot of the experimental data for $A_{r,i}$ vs.R, and $A_{r,f}$ vs. R, where, $R = R(r)$ is a suitable function of r, $(0 \le r \le \infty)$, to be termed as **R-function.**

Thus, in the selfing equation, (apart from the two setting times t_i and t_f as known), $A_{r,t}$ is expressed in terms of $A_{r,i}$, $A_{r,f}$, and t, and again, both $A_{r,i}$ and $A_{r,f}$ are expressed in terms of r.

3.3 REPRESENTATIONS OF THE SELFING EQUATION

The selfing equation (3.4) or its alternative forms (3.5 or 3.6) can be represented in several ways. As stated earlier **(Art 2.2)**, there are **six quantities** to handle in selfing or crossing problems. Here in the A_r, - equation, they are:

t_i and t_f	: **known** (once the mix A is decided)
$\mathbf{A}_{r,i}$ and $\mathbf{A}_{r,f}$	: **known** (the corresponding strengths at these two setting times, once the particular variation r is decided)
t	: **known** (once the time lag is decided)
$A_{r,t}$	: **to be found** (the strength at this time lag t).

That is, **out of six quantities, if five are known, the sixth can be found out.**

The following representations of the selfing equation are aimed at obtaining the same goal in various alternative forms with the understanding that, the r-variation is arbitrary; $(0 \le r \le \infty)$.

1. Graphical (or Vector) representation: Given the strengths $A_{r,i}$ at t_i and $A_{r,f}$ at t_f, to find $A_{r,t}$ at any time lag t.

Use (3.5), that is

$$\frac{t_f - t_i}{A_{r,t}} = \frac{t - t_i}{A_{r,f}} + \frac{t_f - t}{A_{r,i}}$$

Construction: Refer to **Fig. 3.2.**

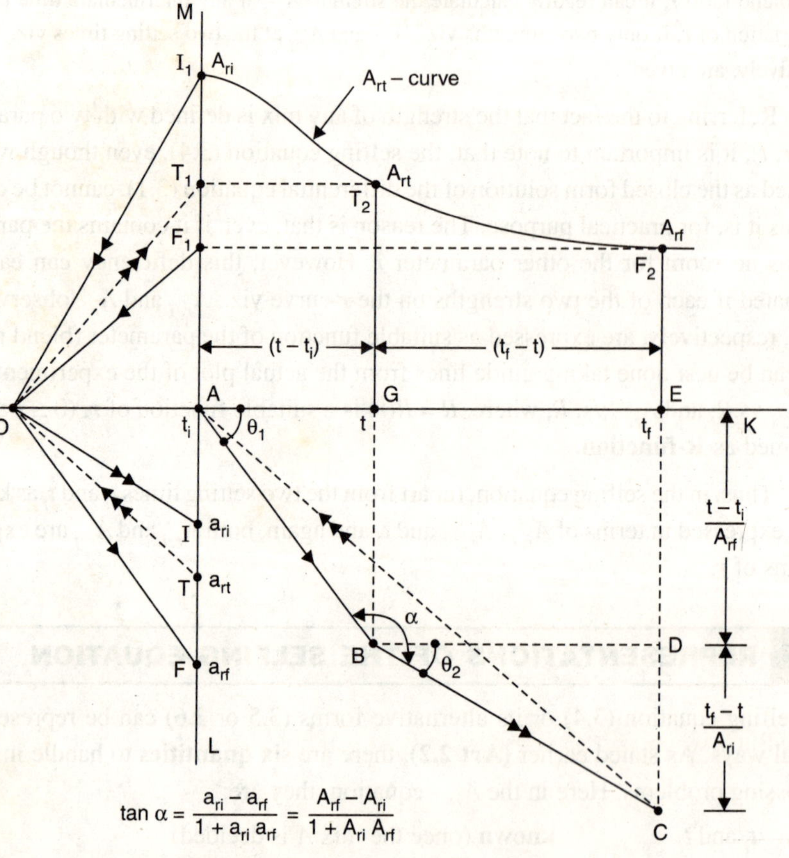

Fig. 3.2. Graphical representation of the selfing equation

Draw line AK on which locate A for $t = t_i$, E for $t = t_f$, and G for t.

Draw vertical ordinates both upwards (for strengths A_{rt}) and downwards (for reciprocal of strengths $a_{rt} = 1/A_{rt}$), through A, G and E.

Mark I_1 for A_{ri} and F_1 for A_{rf} on *AM* so that the strength-time curve $I_1T_2F_2$ can be fully drawn.

Extend line *KA* up to *O* such that $OA = 1$ (unity).

To get the reciprocal of the strengths a_{ri} and a_{rf}, join I_1O and F_1O, and draw perpendicular lines *IO* and *FO* respectively, to get *I* and *F* on line *AL*.

Draw *AB* parallel to *OF* to get point *B* on the vertical ordinate at point *G* at location of *t*, and *BC* parallel to *OI* to get point *C* on *EC*.

Join *AC*, and from *O* draw *OT* parallel to it to locate *T* on *AL*, to represent $a_{rt} = 1/A_{rt}$ on *AL*.

Draw OT_1 perpendicular to *TO* to get point T_1 to represent the required strength $AT_1 = GT_2 = A_{rt}$.

Proof: $AG = t - t_i$, $GE = t_f - t$, $AE = t_f - t_i$.

Since ΔI_1OA and ΔIOA are similar, and $OA = 1$,

$$\frac{AI}{OA} = \frac{OA}{AI_1} \quad \Rightarrow \quad AI = \frac{1}{AI_1} = \frac{1}{A_{r,i}} = a_{ri}$$

Since ΔF_1OA and ΔFOA are similar, and $OA = 1$,

$$\frac{AF}{OA} = \frac{OA}{AF_1} \quad \Rightarrow \quad AF = \frac{1}{AF_1} = \frac{1}{A_{rf}} = a_{rf}$$

Since ΔOAF and ΔAGB are similar, and OA = 1,

$$\frac{BG}{AF} = \frac{AG}{OA} \quad \Rightarrow \quad BG = DE = AF.AG = a_{rf}.(t - t_i) = \frac{t - t_i}{A_{rf}}$$

Since ΔOAI and ΔBDC are similar, and $OA = 1$,

$$\frac{CD}{IA} = \frac{GE}{OA} \quad \Rightarrow \quad CD = IA.GE = a_{ri}.(t_f - t) = \frac{t_f - t}{A_{ri}}$$

Adding, $$CE = CD + DE = \frac{t_f - t}{A_{ri}} + \frac{t - t_i}{A_{rf}}$$

But by virtue of the selfing expression (3.5), it is equal to $\dfrac{t_f - t_i}{A_{rt}}$

Since ΔOAT and ΔAEC are similar, and $OA = 1$,

$$\frac{AT}{OA} = \frac{CE}{AE} \quad \Rightarrow \quad AT = \frac{CE}{AE} = \frac{CE}{t_f - t_i} = \frac{1}{A_{rt}} = a_{rt}$$

Thus $AT_1 = \dfrac{1}{AT} = A_{rt}$ is the required ordinate obtained graphically.

Since ΔABC is a closed triangle, it can be analogously inferred that, the resultant of the forces AB and BC is the force AC, which when represented vectorially, appears

$$(t - t_i).\ \bar{a}_{rf} + (t_f - t).\ \bar{a}_{ri} = (t_f - t_i)\ .\ \bar{a}_{rt} \tag{3.7}$$

where, **the reciprocal of the strengths *i.e.*, a_{rt} (= $1/A_{rt}$), are the corresponding force vectors, and the time lag quantities are the corresponding scalar multipliers.**

Further from **Fig. 3.2**, and noting that, $\alpha = \pi - (\theta_1 - \theta_2)$

where, $\angle GAB = \theta_1$, and $\angle DBC = \theta_2$.

$$\tan \theta_1 = \frac{(t - t_i)\,.\,a_{rf}}{t - t_i} = a_{rf}, \quad \text{and} \quad \tan \theta_2 = \frac{(t_f - t)\,.\,a_{ri}}{t_f - t} = a_{ri} \tag{3.8}$$

$$\tan \alpha = -\tan(\theta_1 - \theta_2) = -\frac{\tan \theta_1 - \tan \theta_2}{1 + \tan \theta_1 \tan \theta_2} = \frac{a_{ri} - a_{rf}}{1 + a_{ri}\, a_{rf}} = \frac{A_{rf} - A_{ri}}{1 + A_{ri}\, A_{rf}} \tag{3.9}$$

For equivalent strength: Following the same construction as described in **Fig. 3.2**, the graphical construction for the case of equivalent strengths is given in **Fig. 3.3**.

Points N and U are for $A\ (r_1, t_1) = AN$, and $A(r_1, t_f) = EU = AX$ for r_1 – curve, and points M and V are for $A(r_2, t_i) = AM$, and $A(r_2, t_f) = EV = AY$ for r_2 – curve.

Given that, $A\ (r_1, t_1)$ and $A\ (r_2, t_2)$ are equivalent at time lags t_1 and t_2, with blend ratios r_1 and r_2 respectively, *i.e.*, $SQ = TR = AP$.

Then viewing vectorially, the resultant of the forces AB and BC, and that of the forces AD and DC will be the same force AC.

In view of the strength equivalence $A\ (r_1, t_1) = A\ (r_2, t_2)$, and referring to (3.9), it is interesting to conclude that, if

$\dfrac{A_{ri} - A_{rf}}{1 + A_{ri}\, A_{rf}}$**, for $r = r_1$ and $r = r_2$, are identical (which is the case for $\angle ABC = \angle ADC$), the peak points B and D (for r_1 and r_2 cases respectively), would lie on the circular arc $ABDC$ drawn on the chord AC.**

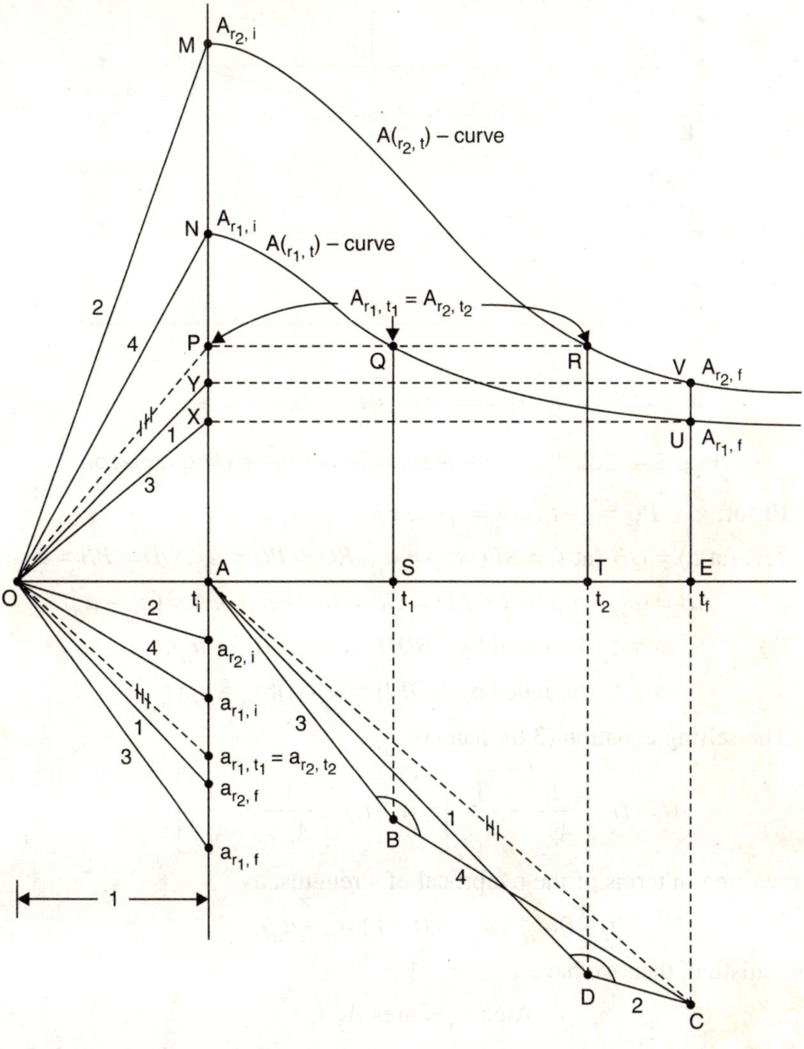

Fig. 3.3. Graphical representation of equivalent strength $A(r_1, t_1)$ and $A(r_2, t_2)$

2. Equal area representation: Given the strengths $A_{r,\,i}$ at t_i and $A_{r,\,f}$ at t_f, to find $A_{r,\,t}$ at any time lag t.

In **Fig. 3.4**, the reciprocal of the strengths $\dfrac{1}{A_{rf}} = a_{rf}$ and $\dfrac{1}{A_{ri}} = a_{ri}$, have been located on the vertical axis from P as PA and PC, respectively. On the horizontal axis, the corresponding time lags t_i and t_f are located at P and R, respectively.

Corresponding to time lag t at Q, the reciprocal of the required strength, namely $\dfrac{1}{A_{rt}} = a_{rt}$, will be QD measured from Q.

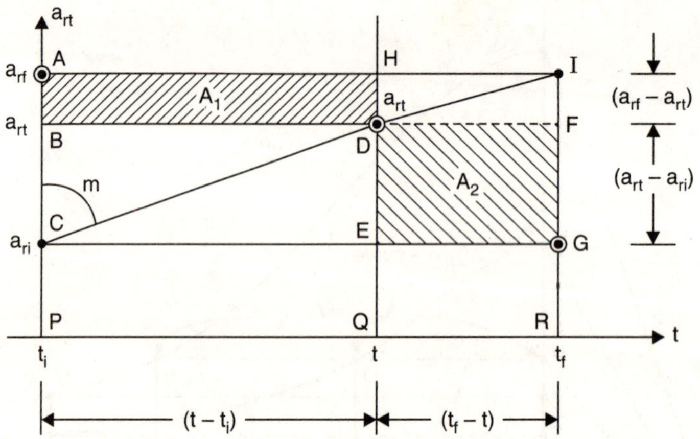

Fig. 3.4. Equal area representation of the selfing equation

Proof: $PQ = t - t_i$, $QR = t_f - t$, $PR = t_f - t_i$

PA (at t_i) = QH (at t) = RI (at t_f) = a_{rf}, $RG = PC = a_{ri}$, $QD = PB = a_{rt}$

$AB = (a_{rf} - a_{rt})$, $CB = ED = GF = (a_{rt} - a_{ri})$, $CA = (a_{rf} - a_{ri})$

So, area A_1 (bounded by $ABDH$) = $(t - t_i)(a_{rf} - a_{rt})$,

and, area A_2 (bounded by $HDFI$) = $(t_f - t)(a_{rt} - a_{ri})$

The selfing equation (3.6), namely

$$(t_f - t)\,.\left[\frac{1}{A_{r,t}} - \frac{1}{A_{r,i}}\right] = (t - t_i)\,.\left[\frac{1}{A_{r,f}} - \frac{1}{A_{r,t}}\right]$$

when rewritten in terms of the reciprocal of strengths, as

$$(t_f - t)(a_{rt} - a_{ri}) = (t - t_i)(a_{rf} - a_{rt}) \tag{3.10}$$

is to be satisfied, then we have,

$$\text{Area } A_1 = \text{area } A_2 \tag{3.11a}$$

This relation suggests that, the magnitude of the reciprocal of the selfed strength should be such that, the vertical line through any time lag t lying between t_i and t_f, makes two areas between the horizontal lines through a_{rf} and a_{rt}, and through a_{rt} and a_{ri}, equal.

[This statement may be considered to be analogous to the location of the neutral axis in the beam cross section having at least one axis of symmetry, subjected to fully plastic moment.]

$t - a_{rt}$ pair form for selfing equation

The previous statement immediately suggests another alternative method of calculating value of A_{rt} corresponding to the given time lag t, or vice versa.

Refer to **Fig. 3.4** once again with the following definitions as:

- a vertical line through any time lag t as: ***t* – line**
- a horizontal line through any a_{rt} (*i.e.*, reciprocal of the strength A_{rt}): as $\boldsymbol{a_{rt}}$ **– line**
- a pair of mutually perpendicular t – line and its corresponding a_{rt} – line as: $\boldsymbol{t - a_{rt}}$ **pair**
- a rectangle formed by a pair of boundary t – lines and a_{rt} – lines as: $\boldsymbol{t - a_{rt}}$ **rectangle**, and its diagonal as: $\boldsymbol{t - a_{rt}}$ **diagonal**.

Then, the rectangle *CAIG* formed by the pair of boundary lines through a_{ri} and a_{rf}, and t_i and t_f, is divided equally into two triangular areas *CIG* and *CIA* by its diagonal *CI*.

Join *CD* for rectangle *CBDE*, and *DI* for rectangle *DHIF*.

Now since *CD* is the diagonal of the rectangle *CBDE*, area *CDB* = area *CDE*, and since *DI* is the diagonal of the rectangle *DHIF*, area *DIH* = area *DIF*.

Hence, area of rectangle *ABDH* (= A_1) = area of rectangle *DFGE* (= A_2), which is the case as proved earlier (3.11*a*).

This observation concludes that the point *D* must lie on the diagonal *CI*.

Thus it can be inferred that, through any point on the diagonal of the rectangle formed by pairs of extreme *t* – lines and $\boldsymbol{a_{rt}}$ – lines, if a $\boldsymbol{t - a_{rt}}$ pair is drawn for a any particular blend ratio *r*, they will represent reciprocal of the strength $\boldsymbol{a_{rt}}$ corresponding to the given time lag *t*, or vice versa.

This form of graphical construction, viz., $\boldsymbol{t - a_{rt}}$ **pair form, by far, is the simplest representation of the selfing equation.**

Equation of the $\boldsymbol{t - a_{rt}}$ diagonal

In **Fig. 3.4** the $\boldsymbol{t - a_{rt}}$ **diagonal** *CI* is a straight line and its equation is given by

$$a_{rt} = a_{ri} + \frac{a_{rf} - a_{ri}}{t_f - t_i}\,(t - t_i)$$

This can be shown as a direct consequence of the selfing equation (3.4) when suitably written in terms of the reciprocal of the strengths.

Recall (3.4) and divide the numerator as well as the denominator of the RHS by $(A_{ri}\,A_{rf})$ to get

$$A_{rt} = \frac{t_f - t_i}{t(a_{rf} - a_{ri}) + t_f a_{ri} - t_i a_{rf}}$$

$$\Rightarrow \qquad (t_f - t_i)\,a_{rt} = t\,(a_{rf} - a_{ri}) + t_f\,a_{ri} - t_i\,a_{rf}$$

Adding and subtracting $(t_i\,a_{ri})$ to the *RHS* of this equation and rearranging, and finally dividing both sides by $(t_f - t_i)$, we get the same equation of the $\boldsymbol{t - a_{rt}}$ **diagonal**.

Note: We know that, if $y = f(x)$ is linear, (as the $t - a_{rt}$ diagonal is in this case), $y = 1/f(x)$ will be hyperbolic. So the trend of the family of the strength – time curves as given by the selfing equation (3.4) indicates that, it conforms to hyperbolic for the analytical treatment considered.

Alternative expressions of (3.10)

Alternatively, (3.10) in the light of (3.11*a*) can be viewed as follows:

Since the two rectangles are of same area, (that is, area A_1 = area A_2),

$$(t_f - t)(a_{rt} - a_{ri}) = (t - t_i)(a_{rf} - a_{rt})$$

$$\Rightarrow \qquad \frac{a_{rf} - a_{rt}}{a_{rt} - a_{ri}} = \frac{t_f - t}{t - t_i} \tag{3.11b}$$

$$\Rightarrow \qquad \frac{a_{rf} - a_{ri}}{a_{rf} - 2a_{rt} + a_{ri}} = \frac{t_f - t_i}{t_f - 2t + t_i}$$

Substituting, $\bar{t} = (t_i + t_f)/2$ = constant,

and $\bar{a}_{rt} = (a_{ri} + a_{rf})/2$ = constant,

and $\beta = AI/AC$ (inclination of the diagonal CI measured from the a_{rt} – line at t_i)

$= (t_f - t_i)/(a_{rf} - a_{ri})$ = constant,

in this equation, and rearranging, we get

$$\frac{\bar{t} - t}{\bar{a}_{rt} - a_{rt}} = \frac{t_f - t_i}{a_{rf} - a_{ri}} = \beta \text{ (= constant)}$$

Then we have

$$a_{rt}\beta - t = \bar{a}_{rt}\beta - \bar{t} = \text{constant } C$$

So, $$a_{rt}\beta - t = C \tag{3.11c}$$

Any one of (3.10), (3.11*b*) or (3.11*c*) can be used as the alternative forms of equal area representation of the selfing equation.

$t - a_{rt}$ form of representation of equivalent strength and equivalent time

The representation of the equivalent strength and its corresponding equivalent time using $t - a_{rt}$ form is described in **Fig. 3.4A.**

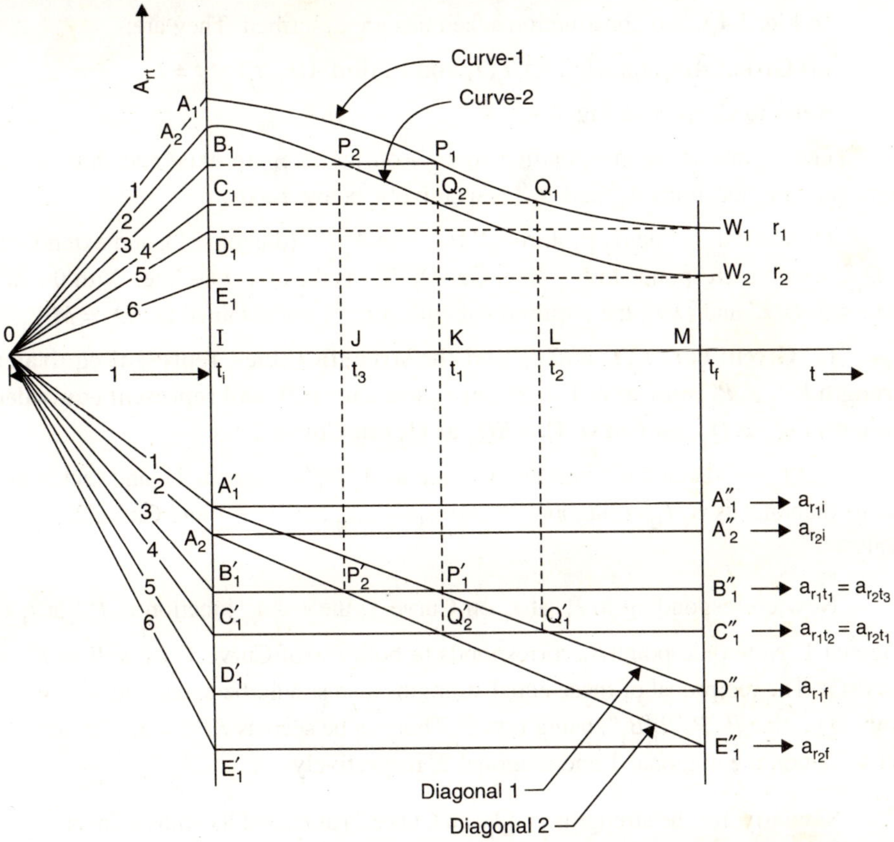

Fig. 3.4*A*. Equivalent strength-time relation through $t - a_{rt}$ pair method

Curve 1 (for $r = r_1$) and Curve 2 (for $r = r_2$) are shown above t-axis. On Curve 1, A_1, P_1, Q_1 and W_1 represent the strengths at t_i, t_1, t_2 and t_f, respectively, measured from t-axis. Similarly, A_2, P_2, Q_2 and W_2 are the corresponding ones at t_i, t_3, t_1 and t_f, respectively. The reciprocal of the strengths are drawn below t-axis.

The rays drawn from O (unit distance from I) on the strength axis are marked as 1, 2, ..., 6, and the reciprocal strength values are again marked by the same sequence of numbering viz., 1, 2, ..., 6. That is, for example, both OA_1 and OA_1' (perpendicular to each other) are marked as 1.

The junction of the $t - a_{rt}$ pairs are: A_1', P_1', Q_1' and D_1'' representing $A(r_1, t_i)$, $A(r_1, t_1)$, $A(r_1, t_2)$ and $A(r_1, t_f)$, respectively, all lie on the $t - a_{rt}$ diagonal $A_1'D_1''$ (diagonal 1) for the $t - a_{rt}$ rectangle $A_1'A_1''D_1''D_1'$.

Similarly, P_2', Q_2' and E_2'', all lying on the diagonal 2, and representing P_2 for $A(r_2, t_3)$, Q_2 for $A(r_2, t_1)$ and E_2'' for $A(r_2, t_f)$, respectively, are for the $t - a_{rt}$ rectangle $A_2'A_2''E_2''E_1'$.

In **Fig. 3.4*A*** two construction schedules are described. They are:

(*a*) Given: $A(r_1, t_i)$, $A(r_1, t_1)$, t_i, t_f and t_1, find $A(r_1, t_f)$ at $t = t_f$

Refer to Curve 1 in **Fig. 3.4*A***.

Get A_1 and B_1 (corresponding to P_1) on the strength axis, and thus get the corresponding locations A_1' and B_1', respectively, below *t*-axis.

Draw the a_{rt} – line $B_1'P_1'$ through P_1'. Join $A_1'P_1'$ (diagonal 1), and extend it up to D_1'', which corresponds to t_f – line. Get $D_1''D_1'$ as the $a(r_1, t_f)$ – line. With the help of rays 5, OD_1' and OD_1, the required strength A (r_1, t_f) is obtained as $ID_1 = MW_1$.

(*b*) Given: t_i, t_f, t_1, t_2 and t_3, and the strength values, represent equivalent strength KP_1 at P_1 (on Curve 1) = JP_2 at P_2 (on Curve 2), and represent equivalent strength LQ_1 at Q_1 (on Curve 1) = KQ_2 at Q_2 (on Curve 2)

Get $t - a_{rt}$ rectangle 1 and its diagonal as $A_1'D_1''$ (diagonal 1), and rectangle 2 and its diagonal as $B_1'E_2''$ (diagonal 2) corresponding to Curve 1 and Curve 2, respectively.

Now corresponding to P_1 at t_1 on Curve 1, the $t - a_{rt}$ junction is P_1' at t_1 on diagonal 1. Note that, point B_1 corresponds to both P_1 (of Curve 1) as well as P_2 (of Curve 2). The reciprocal of these equal strengths also should be equal and is obtained by the a_{rt} – line $B_1'P_2'P_1'B_1''$, using rays 3. This can be seen as $KP_1' = JP_2'$, where P_1' and P_2' lie on the diagonal 1 and diagonal 2, respectively.

Similarly, for the strengths at Q_1 (on Curve 1) at t_2, and its equivalent at Q_2 (on Curve 2) at t_1, represented by point C_1, provides the a_{rt} – line as $C_1'C_1''$ on which Q_1' and Q_2' lie. The former is on the diagonal 1 and the latter on the diagonal 2.

In **Fig. 3.4*B*** the comparison of the **vector form** and ***t* – *a*$_{rt}$ form** of representation of the equivalent strength A (r_1, t_1) at P on Curve 1 and A (r_2, t_2) at Q on Curve 2 is described.

Two points P and Q representing $A(r_1, t_1)$ at P on Curve 1, and $A(r_2, t_2)$ at Q on Curve 2, are strength wise same. The graphical construction of the rays from O (on the *t*-axis, and at unit distance from *K*) are marked as 1*A*, 2*A*...5*A* (for the points denoting strengths above *t*-axis with the letter *A*), and as 1*a*, 2*a* ... 5*a* (for the points denoting the corresponding reciprocal of strengths below *t*-axis with the letter *a*). Thus the $t - a_{rt}$ rectangles are *CDEF* (rectangle 1) and *GHIJ* (rectangle 2) with their respective diagonals *CE* and *GI* for Curve 1 and Curve 2, respectively.

Following the description given in **Fig. 3.4*A***, get points P_1 at the junction of *XY* with t_1 – line on diagonal 1, and Q_1 at the junction of *XY* with t_2 – line on diagonal 2, respectively.

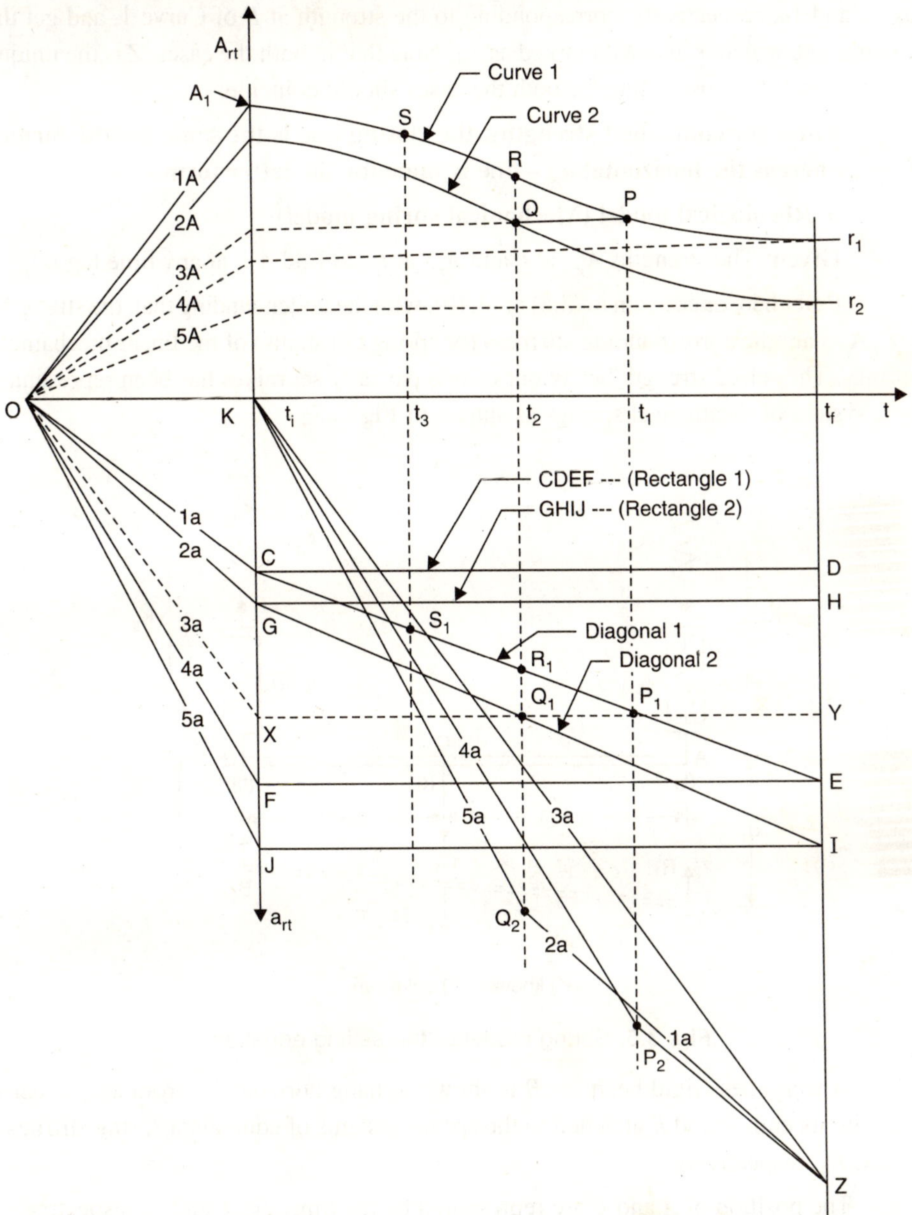

Fig. 3.4*B*. Equivalent strength using vector and $t - a_{rt}$ representation for $A_{r_1 t_1}$ and $A_{r_2 t_2}$

Following the description given in **Fig. 3.3**, get KQ_2, Q_2Z as the parallel lines marked as $5a$ and $2a$, respectively, corresponding to the strength at Q of Curve 2. The closing line is KZ marked as $3a$. Similarly, get KP_2 and P_2Z as the parallel lines marked as $4a$ and $1a$, respectively, corresponding to the strength at P of Curve 1, and get the closing line which is also KZ marked as $3a$. Note that in both the cases, Z is the unique point so that the closing line for both the cases should coincide.

Thus, for equivalent strengths, the closing line is the same for the former form, whereas the horizontal a_{rt} – line is same for the latter form.

3. Rheological model (Mechanical spring model)

Given: The strengths $A_{r,i}$ at t_i and $A_{r,f}$ at t_f, to find $A_{r,t}$ at any time lag t.

Refer here once again to (3.6) or (3.10), with the understanding that, the strengths viz., A_{ij}-s are the corresponding stiffness (or spring constants) of the linear mechanical springs. The selfed strength behaviour of two partially set mixes has been represented by a system of mechanical springs as shown in **Fig. 3.5**.

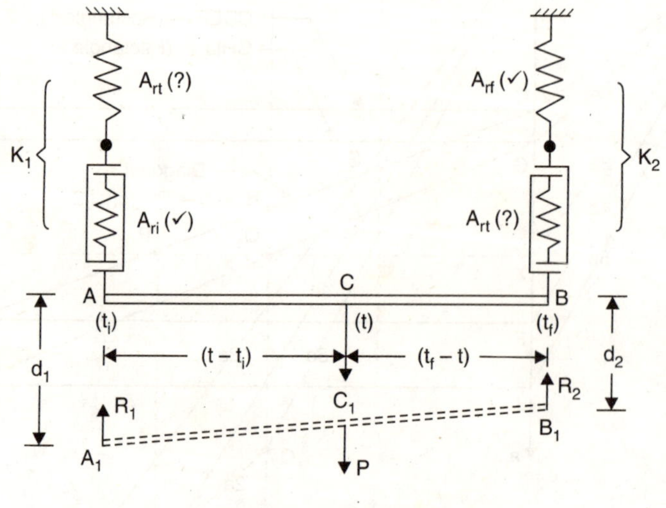

Fig. 3.5. Spring model of the selfing equation

A weightless rigid beam ACB is shown to hang horizontally from a rigid ceiling, with its ends A and B attached to the spring systems of equivalent spring stiffness K_1 and K_2, respectively.

The position of A and B are represented by the time lags t_i and t_f, respectively, so that the length of AB is constant and represents the time lag difference $(t_f - t_i)$. The position of C is a variable point representing the time lag t, at which the selfed strength is intended to be found. Thus, $AC = (t - t_i)$, and $CB = (t_f - t)$.

Let us now define: **a spring to be positive, when it elongates under an applied tensile force, or when it shortens under an applied compressive force, and negative, when it shortens under an applied tensile force, or when it elongates under an applied compressive force.**

The left system K_1 is the equivalent spring constant of the positive spring A_{rt} (**to be found**) and negative spring A_{ri} (**given**) arranged in series, and the right system K_2 is the equivalent spring constant of the negative spring A_{rt} (**to be found**) and positive spring A_{rf} (**given**) arranged in series. The entire spring system consists of K_1 and K_2 arranged in parallel, in which spring stiffnesses A_{ri} and A_{rf} are the **given** ones, and stiffness A_{rt} is **to be found.**

Since each of the spring system K_1 as well as K_2 is composed of one positive and negative spring in series, the expression for the resultant spring stiffness is written as

$$\frac{1}{K_1} = \left[\frac{1}{A_{rt}} - \frac{1}{A_{ri}}\right] = (a_{rt} - a_{ri}), \text{ and } \frac{1}{K_2} = \left[\frac{1}{A_{rf}} - \frac{1}{A_{rt}}\right] = (a_{rf} - a_{rt}) \tag{3.12}$$

Consider that a concentrated load P is applied at C. The end reactions are R_1 at A and R_2 at B.

$$R_1 = P.\frac{BC}{AB} = P.\frac{t_f - t}{t_f - t_i}$$

$$R_2 = P.\frac{AC}{AB} = P.\frac{t - t_i}{t_f - t_i} \tag{3.13}$$

These end reactions cause displacements d_1 and d_2 at A and B, respectively. The displaced position of AB is A_1B_1 (shown dotted).

Since, displacement (d) = reaction (R)/stiffness (K), we have with the help of (3.12) and (3.13)

$$d_1 = \frac{R_1}{K_1} = P.\frac{t_f - t}{t_f - t_i}.\left[\frac{1}{A_{rt}} - \frac{1}{A_{ri}}\right] = P.\frac{t_f - t}{t_f - t_i}.(a_{rt} - a_{ri})$$

$$d_2 = \frac{R_2}{K_2} = P.\frac{t - t_i}{t_f - t_i}.\left[\frac{1}{A_{rf}} - \frac{1}{A_{rt}}\right] = P.\frac{t - t_i}{t_f - t_i}.(a_{rf} - a_{rt}) \tag{3.14}$$

In general, d_1 and d_2 are of different magnitudes, and as a result, the deflected position of the beam A_1B_1 is tilted.

But in the light of the selfing equation (3.6) or (3.10), (3.14) yields the **equality**

$$d_1 = d_2 \tag{3.15}$$

This statement ensures that, **the value of the spring stiffness A_{rt} (both simultaneously in the spring systems K_1 and K_2) is the right one for which the beam *ACB*, acted on by a load at the variable position of time lag *t* at *C*, becomes horizontal. In other words, the beam only displaces downwards without rotation.**

4. Plate deflection analogy

The ideal trend of the strengths $A_{rt} = A(r, t)$ of the blended partially set concretes with respect to the time lag t for different blend ratios r; $(0 \le r \le \infty)$, can be observed to be in a very close agreement to the trend of the deflection contours $w = w(\bar{x}, \bar{y})$, plotted along $\bar{x}$ $(= x/a)$, for different $\bar{y}$ $(= 2y/b)$ locations, of a rectangular thin plate $(a \times b)$ having a particular type of boundary conditions (viz., simply supported along three edges $\bar{x} = 0$, $\bar{y} = 0$ and $\bar{y} = 2$, and free along the fourth edge at $\bar{x} = 1$, the origin of $(\bar{x}, \bar{y})$ being at one corner of the plate), under uniform loading.

Typical deflected surface is shown in **Fig. 3.6,** for the half portion of the plate between $\bar{y} = 0$, $\bar{y} = 1$, and $\bar{x} = 0$, $\bar{x} = 1$, considering symmetry of the plate structure and the loading about the line $\bar{y} = 1$.

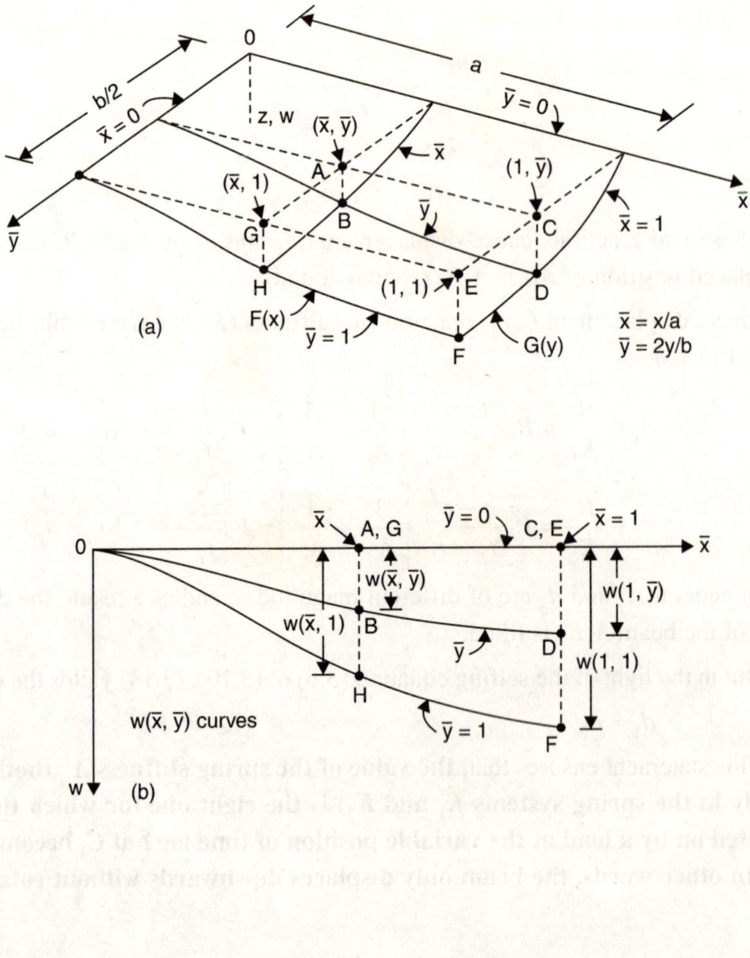

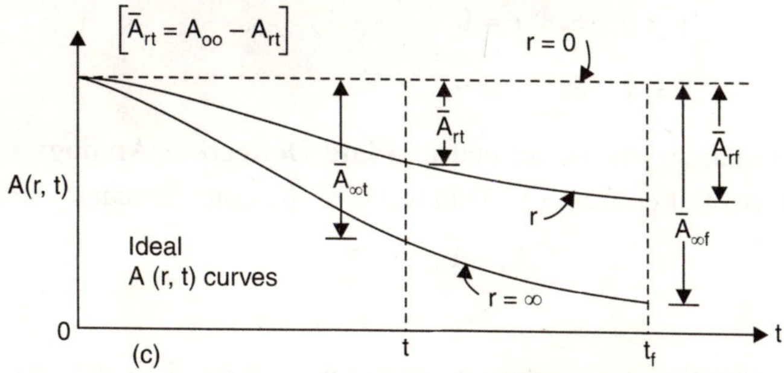

Fig. 3.6. Plate deflection analogy of the selfing equation

Consider the ordinate of the plate deflection AB at any location $(\bar{x}, \bar{y})$ is w $(\bar{x}, \bar{y})$.

Designate the parent deflected curves at $\bar{y} = 1$ (on the symmetry line) as $F(x)$, and at $\bar{x} = 1$ as $G(y)$. Corresponding to the ordinate AB, CD and GH are the two ordinates on $G(y)$ and $F(x)$, respectively, on the parent curves.

It can be checked that, the following approximate geometrical relation connecting displacements holds good, namely

$$AB \times EF = CD \times GH \tag{3.16}$$

which when rewritten becomes

$$w(\bar{x}, \bar{y}) = \frac{w(1, \bar{y}) \,.\, w(\bar{x}, 1)}{w(1, 1)} \tag{3.17}$$

Compare the plate deflection $w(\bar{x}, \bar{y})$ with the idealized strength-time curves A_{rt}, (as in Type *I* variation, **Art. 3.4.1**). It can be noticed that, though the trends of the both are similar, (*i.e.*, drooping downwards towards the increasing values of $\bar{x}$ or t, as the case may be), the strength is measured upwards, whereas the deflection is measured downwards.

So, in view of this, using **complementary strength**, *i.e.*, $\bar{A}_{rt} = A_{oo} - A_{rt}$, and noting the following equivalence, viz.,

(*i*) **deflection** in plate, $w(\bar{x}, \bar{y}) \Leftrightarrow \bar{A}_{rt}$, **complementary strength** in selfing

(*ii*) **distance** in plate, $\bar{x} \Leftrightarrow t$, **time lag** in Selfing

(*iii*) **location** in plate, $\bar{y} \Leftrightarrow r$, **blend ratio** in Selfing

$$\bar{y} = 0 \quad \Leftrightarrow \quad r = 0$$

$$\bar{y} = 1 \quad \Leftrightarrow \quad r = \infty$$

the following analogy (to be termed as **Plate Deflection Analogy of the Selfed Strength**) can be written from (3.17) in terms of the complementary strengths as

$$\bar{A}_{rt} = \frac{\bar{A}_{rf} \cdot \bar{A}_{\infty t}}{\bar{A}_{\infty f}} \tag{3.18}$$

It is to note that, (3.17) does not contain $\bar{y}$ at all. This is reflected in (3.18), where the analogous parameter r (blend ratio) is too absent. In other words, this suggests that, **between any two strength-time curves, one being the basic partially set mix (for which $r = \infty$), and the other of any value of r, (3.18) is applicable.**

3.4 REVIEW AND SUGGESTIONS ON THE SELFING EQUATION AND ITS ASSUMPTIONS

1. Assumption-4 (**Art. 3.1**) is suited for the ideal variation for the strength-time curve where the maximum strength is attended at $t = 0$ for OHC, and all other strengths for THC observed at any time lag t, for any variation of r, lie below this value. But in reality, except for some variety of concretes, this ideal trend is often violated.

Experiments conducted on a wide range of concretes of different variety reveal mainly six types of strength-time variations, namely, Type I, Type II, ..., Type VI, (**Fig. 3.7**).

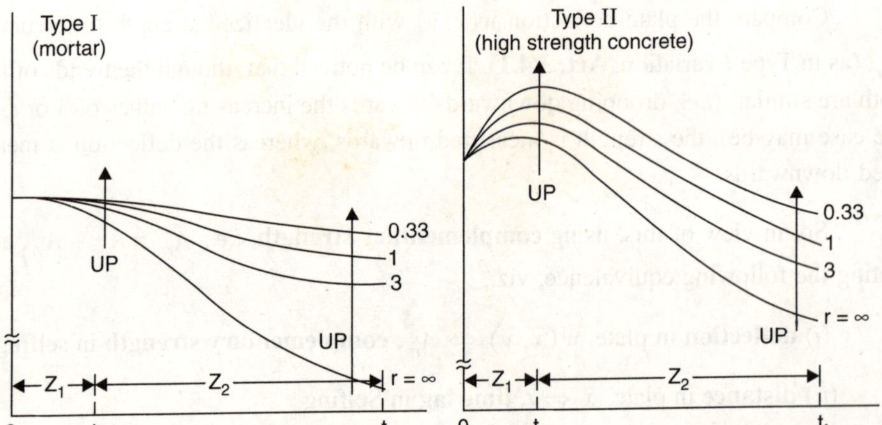

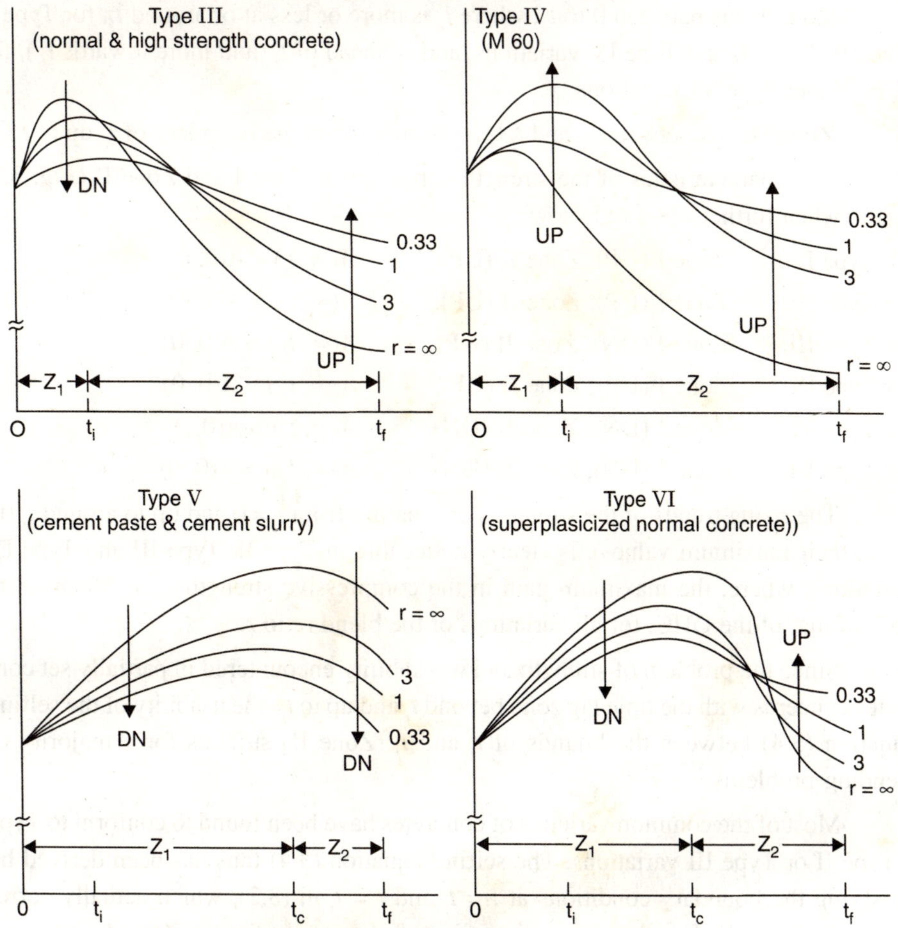

Z = Zone I, Z = Zone II

Fig. 3.7. Types of typical strength-time curves

The time lag zone between $t = 0$ and $t = t_f$, can be broadly divided into two distinct zones, viz., Zone I and Zone II.

The strength-time curves, in each type of variations, within the bound of the two extreme limits of time lags, are seen to have unique value, A_{oo} at $t = 0$, from which a family of curves for different values of r fan out progressively towards the increasing values of t.

Using the terminology viz., UP to mean the basic curve ($r = \infty$) to occupy the bottom most position, and all other curves ($r < \infty$) to lie successively above it, and *DN* to mean the opposite to the UP pattern, the domain of t viz., ($0 \le t \le t_f$) in each of the six types of A_{rt}—variations can be broadly divided into two zones, namely, zone I and Zone II.

Zone I: t is between 0 to t_c, where t_c is more or less at or around t_i, for Type I, Type II, Type III and Type IV variations, and is ahead of t_i (and more towards t_f), for Type V and Type VI variations.

Zone II: t is between t_c and t_f (*i.e.*, starting from the boundary of t_c up to t_f).

The characteristics of the strength variations in Zone I and Zone II (**Fig. 3.7**) again, when further observed show:

for Type I:	Zone I (UP), Zone II (UP),	$A(\infty, t_i) < A(0, 0)$
for Type II:	Zone I (UP), Zone II (UP),	$A(\infty, t_i) > A(0, 0)$
for Type III:	Zone I (DN), Zone II (UP),	$A(\infty, t_i) > A(0, 0)$
for Type IV:	Zone I (UP), Zone II (UP),	$A(\infty, t_i) > A(0, 0)$
for Type V:	Zone I (DN), Zone II (DN),	$A(\infty, t_c) > A(0, 0)$
for Type VI:	Zone I (DN), Zone II (DN),	$A(\infty, t_c) > A(0, 0)$

The rising trends of the strengths originating from $t = 0$ and up to around t_i (to attain their maximum values), is clearly noticeable in Type II, Type III and Type IV variations, where, the maximum gain in the compressive strength f_c, is about 18 to 25% of that of the OHC, for all variations of the blend ratio r.

Since the problem of strength and workability encountered in partially set concretes concerns with the time lag zone beyond t_i and up to t_f, the usability of the selfing equation (3.4) between the bounds of t_i and t_f (Zone II) suffices for a majority of blending problems.

Most of the common varieties of concretes have been found to conform to Type I, Type II or Type III variations. The selfing equation (3.4) thus has been derived by satisfying the boundary conditions at $t = t_i$ and $t = t_f$ in (3.2), which actually corresponds to Zone II. For the appropriate formula corresponding to Zone I, the same equation (3.4) can be directly used on replacing t_i and t_f by 0 and t_i respectively.

Thus, the selfing equation becomes:

for **Zone I:** $$A_{r,t} = \frac{A_{r,o} \cdot A_{r,i} \cdot (t_i)}{t \cdot A_{r,o} + A_{r,i} \cdot (t_i - t)}$$

for **Zone II:** $$A_{r,t} = \frac{A_{r,i} \cdot A_{r,f} \cdot (t_f - t_i)}{t \cdot (A_{r,i} - A_{r,f}) + A_{r,f} \cdot t_f - A_{r,i} \cdot t_i} \tag{3.19}$$

However, to apply selfing equation in Type V and Type VI variations, no special problem is encountered at all, and in those cases, t_i is to be replaced suitably by t_c in the said formula.

The prediction of the blended strength of a mix, on the basis of the Assumption-4, has been found to be sufficiently accurate. This is particularly true when Zone I is comparatively shorter than the Zone II, and further, when the difference between the strengths corresponding to the two setting times is small. But even then Assumption-5 cannot be said to hold true for a curve whatever flat it may be.

2. The constant slope *m* for each variation of *r* in the family of the strength-time variations, conforming to Assumption-5, is strictly true, if each variation in the said family of the curves is linear. However, this is not the case. The linear form can be achieved with reasonable and acceptable accuracy, if the strengths are plotted against **time lag parameter** *T*, instead of time lag *t*, on describing *T* as a suitable function of *t*, such as, $T = t^n$, where, n = an integer.

   Thus, with a suitable value of *n*, corresponding to a particular strength for the given mix, the *A*(*r*, *T*)–variations suffice the requirement for the mathematical treatment for the selfed strength.

   Whether the strength vs time lag variation is taken as family of linear lines or not, the analytical results appear to be too close to differentiate. (A practical numerical problem solved in Example 4.6 strengthens this view).

   In **Fig. 3.8**, a typical plot of A_{rt} and its corresponding A_{rT} are shown.

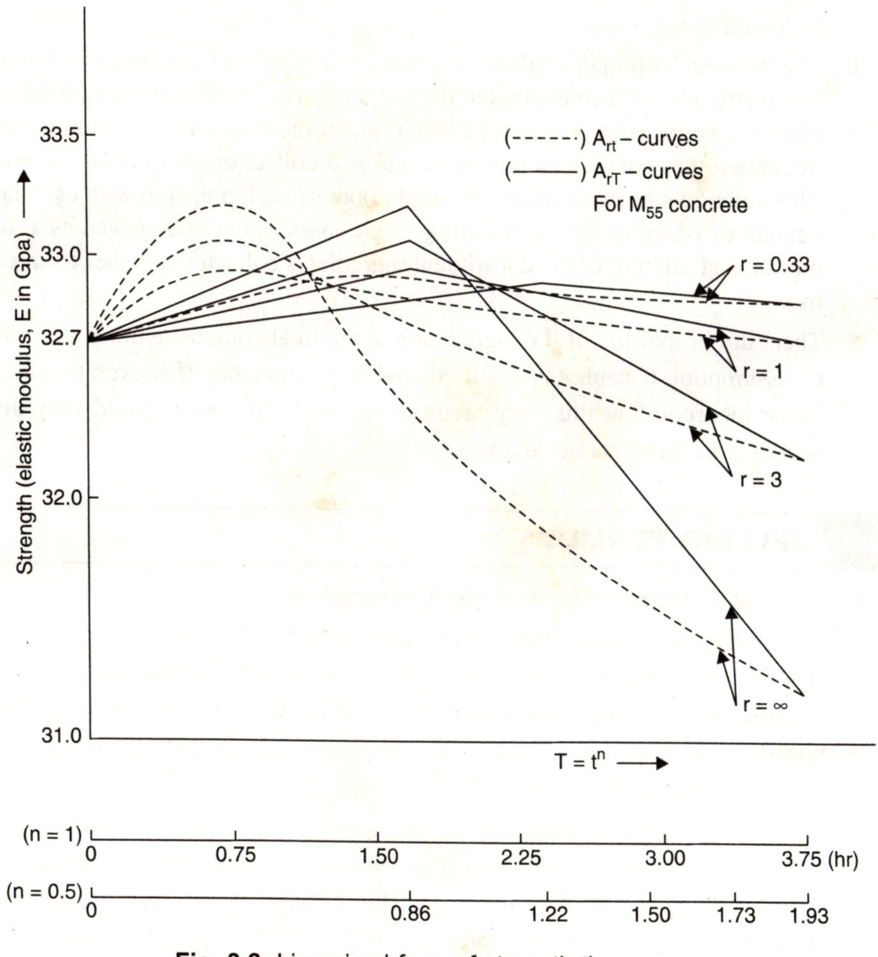

Fig. 3.8. Linearised form of strength-time curves

3. Since, the degree of partial setting varies with time lag of exposure in air for concretes, the required time period of vibration during casting into the mould too does vary, and in no case, be same. Then, to ensure maximum strength from partially set concretes, the duration of the compaction needs to be designed, and to be applied through out same. For individual mixes, the duration of the vibration time for individual mixes should be so decided as to get each sample at different time lags just compacted leaving no void and air bubble inside the mass. This may be ensured from the appearance of oozing out of water film on the sample in the mould. Obviously, the compaction time at different time lags is bound to be different. More is the degree of partial setting; more is the requirement of compaction time.

   However, during selfing of two partially set mixes (each considered at different time lags), the full compaction cannot be so easily designed. In the absence of any guide lines, just oozing out of water film on the selfed sample in the mould may be accepted as the sufficient and right compaction, and the same may be followed through out.

4. The presence of proper quantity of water in a concrete mix is aimed at designed target strength. In a concrete, the ingredients (viz., cement, fine aggregate and coarse aggregate) in presence of water, go in the reaction producing a single integrated mass, which cannot be taken as a collection of isolated materials. Thus, blending of two such reactive mix bodies, each prepared with calculated amount of water, to form a resulting single mass, cannot be viewed as a mere mechanical mixture of the constituent ingredients collected together from each mix.

   Then, dry proportion of the ingredients in the final composite mass, according to Assumption-6, cannot be justified, and is questionable. However, the formulation incorporating this very assumption has been seen to yield very good results with sufficient accuracy.

3.5 SELFING FORMULA

1. Concept of characteristic strength parameters

The Selfing equation (3.4) is valid for any value of r; $(0 \leq r \leq \infty)$, and t; $(0 \leq t \leq t_f)$. However, this equation cannot be used as it is for solving practical problems. This is because of the fact that, in calculating blended strength A_{rt}, the right hand side of (3.4), though can accept the value of the time lag t, cannot provide any room to accommodate the value of the blend ratio r.

To overcome this deficiency, express A_{ri} as well as A_{rf} as suitable functions of r, viz., $A_{rk} = R$; (where, $R = f(r)$, and the suffix $k = i$ or f, indicating t_i or t_f, as the case may be), in the form of empirical relations. This is better done using experimental values of the strengths for different values of r; $(0, \ldots \infty)$, observed specifically at two locations viz., at $t = t_i$, and at $t = t_f$.

We discuss the proper selection of the function R as some suitable function of r along the horizontal axis of the plot of the A_{rk} vs R.

The strength values of A_{ri} at t_i, for various values of r, are closer to one another, whereas, the corresponding values of A_{rf} at t_f, are comparatively more separated. For this reason, the trend of A_{rf} against r; ($r = \infty$, ... 0) often appear as curved, whereas, that of A_{ri} against r; ($r = \infty$, ... 0) may often appear as linear. In fact, accepting both of these two variations (viz., A_{ri} vs R, and A_{rf} vs R) as parabolic, the linear form (**Fig. 3.8**) can be readily obtained on replacing the function R by log R.

The linear function is then expressed in slope-intercept form: $Y = m.X + C$, as

$$A_{rk} = C_k + m_k R \tag{3.20}$$

where, the suffix k stands for i or f, representing time lags t_i, and t_f, respectively,

i.e., $$A_{ri} = C_i + m_i .R \text{, and, } A_{rf} = C_f + m_f .R \tag{3.21}$$

(3.20) or (3.21) is to be referred as ***ArkR*-plot** meant for the mix type A, for further reference. Similarly, *BrkR*-plot, *CrkR*-plot apply for mix B, mix C, and so on.

The four quantities viz., intercepts (**C_i** and **C_f**) and slopes (**m_i** and **m_f**), corresponding to A_{ri} as well as A_{rf} variations with respect to R, is termed as **characteristic strength parameters**, or simply, ***CSP-s*** of the mix.

The prediction of the strength of a composite mix prepared out of blending of two partially set mixes, has been made analytically possible through selfing (or crossing) concept, if the characteristic strength parameters as well as the two setting times of each of the individual mixes, are available in hand.

2. Selection of R-function

Depending on the suitable choice of R, along the horizontal axis (**Fig. 3.8A**), the slopes m_k; (m_i or m_f) may be positive or negative, and so also the intercepts C_k; (C_i or C_f), which correspond to the strength values at the origin O.

Remember that, the value ∞ cannot be plotted on the R-axis, and moreover, log (0) and log (∞) remain undefined, the following observations need to be kept in mind while considering the function $R = f(r)$ in **Fig. 3.8A**.

If $R = r$ or, $R = 1/r$, the location corresponding to ∞ or 0, as the case may be, cannot be plotted on the horizontal axis, and hence is naturally discarded. The strength value corresponding to $r = 0$ or $r = \infty$, as the case may be, is the origin O, where the corresponding strength value is the intercept C_k.

For $R = r$, $A_{\infty k}$ values are discarded, and A_{ok} values at the origin O are the C_k, whereas, for $R = 1/r$, A_{ok} values are discarded, and $A_{\infty k}$ values, obtained as extrapolated ones at the origin O, provide C_k.

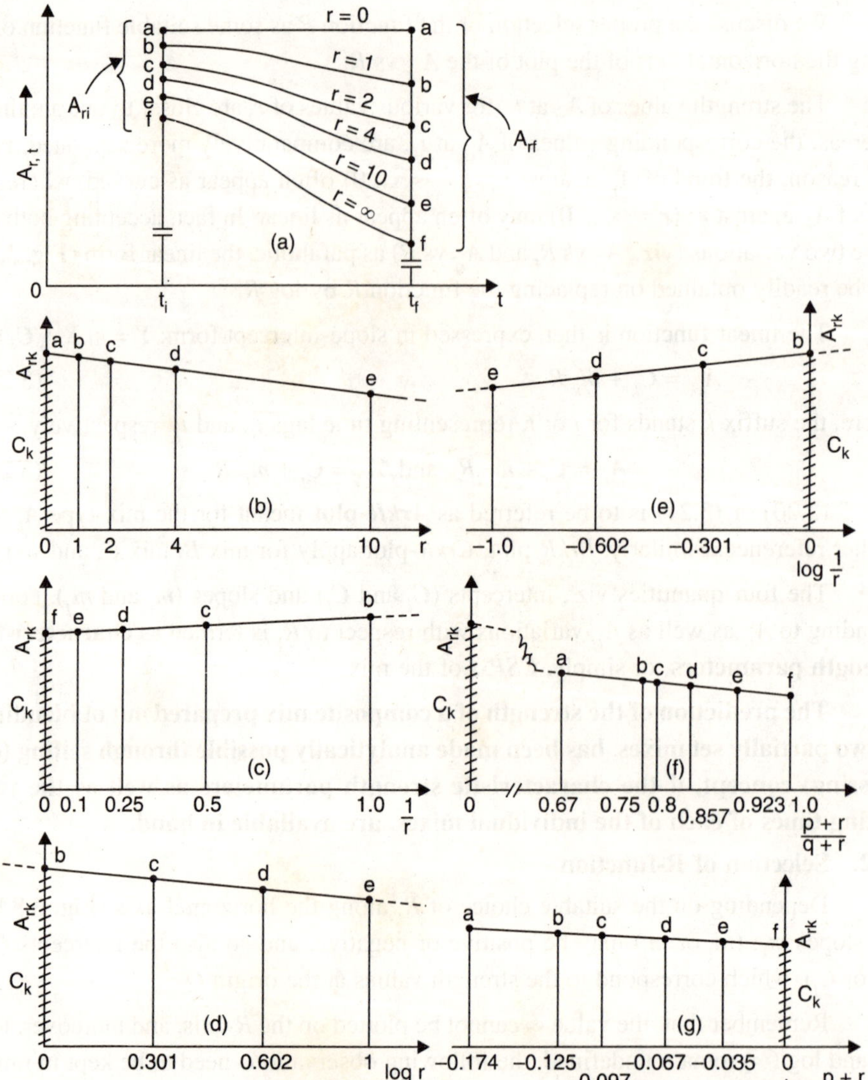

Fig. 3.8 A. $A_{rk}R$ plot ($k = i$ or f)

If $R = \log r$, both the values at $r = 0$ and $r = \infty$, are undefined, and hence, are of no use. The value corresponding to $r = 1$ is the origin O, and those corresponding to r; $(0 < r \le 1)$ and $(1 \le r < \infty)$, remain to the left and to the right to it, respectively. The strength value corresponding to $r = 1$ is the intercept C_k.

If $R = \log 1/r$, all other characteristics remain same as those of the previous case, with the exception that, the values corresponding to r; for $0 < r \le 1$, and for $1 \le r < \infty$, occupy the right and the left of the origin O (at $r = 1$), respectively.

If $$R = \frac{a+r}{b+r}, \qquad \left[\text{so that}, r = \frac{a-b}{R-1} - b\right]$$

or, $$\text{R} = \log \frac{a+r}{b+r}, \quad \left[\text{so that}, r = \frac{a-b}{10^R - 1} - b\right] \qquad (3.22)$$

(where, a and b are integers, and **log** is with base 10),

The difficulty faced in earlier forms of R in locating the value of r corresponding to $r = 0$ and/or $r = \infty$, and for which the corresponding strength values thereof could not be utilized, can be totally eliminated. Thus, at $r = 0$ and $r = \infty$, the first function yields the values a/b and 1, respectively, and the second function yields log (a/b) and 0, respectively, the location of which on the R-axis do not pose any problem, and so the strengths corresponding the two extreme values of r can be used.

Plot of log A_{rk} values (instead of simply, A_{rk}) on the strength axis, against R in log form (3.22), can be attempted to get still better linear form.

The slope–intercept form as: $Y = m \,.\, X + C$, then becomes

$$\log A_{rk} = C_k + m_k .\log R \qquad (3.23)$$

$$= \log D_k + m_k .\log R = \log \left(D_k \,.\, R^{m_k}\right)$$

$$\Rightarrow \quad A_{rk} = D_k \,.R^{m_k} \qquad (3.24)$$

where, D_k = antilog (C_k) = exp(C_k), and, $k = i$ or f (as the case be)

3. Practical form of selfing equation

It is worth to notice that:

(*i*) Suitable relation of A_{rk} as function of r, [for example (3.20) or (3.24)], helps to compute the pair of strengths $A(r, t_i)$ and $A(r, t_f)$ at t_i and t_f, respectively, for any particular value of the blend ratio r lying in the range of 0 to ∞.

(*ii*) Additionally (refer to **Note** in **Art. 3.2**, and recollect that), if the two strengths $A(r_1, t_i)$ and $A(r_1, t_f)$ at t_i and t_f, respectively, only are given, for any particular value of the blend ratio $r = r_1$ (say), the entire strength variation for any value of t, in the given time lag range, corresponding to that blend ratio r_1, is then completely known through any one of (3.4) or (3.5) or (3.6).

Thus, **it is sufficient to know only the two strengths at the two setting times for the basic mix (for which $r = \infty$), and the suitable relations of the strengths at the two setting times only, as functions of r. Then all strengths in the given time lag range and blend ratio range can be computed analytically.**

$A_{r,i}$ and $A_{r,f}$, appearing in the selfing equation (3.4), which already contains t (time lag), now contains r (blend ratio), by virtue of (3.21) or (3.24), in which $A_{r,k}$; ($k = i$ or f), is expressed as function of r.

So finally, in the selfing equation, (apart from the two setting times t_i and t_f, as known), both $A_{r,i}$ and $A_{r,f}$ are expressed in terms of r, and thus, $A_{r,t}$, (which contains $A_{r,i}$ and $A_{r,f}$), is now written in terms of r as well as t.

Thus, substituting (3.21) or (3.24), in (3.4), namely

$$A_{r,t} = \frac{A_{r,i} \, . A_{r,f} \, .(t_f - t_i)}{t.(A_{r,i} - A_{r,f}) + A_{r,f} \, .t_f - A_{r,i} \, .t_i} \tag{3.24a}$$

the selfing equation is now ready for solving practical problems.

4. Resultant blend ratio R′ of two mixes in blending

For general selfing:

Case I: $r_1 \neq r_2$, $t_1 \neq t_2$, observed at $T = t_1$, $(t_1 > t_2)$

Each one of the two partially set mixes are with different blend ratios r, and time lags t, and considered for selfing at the time lag t_1.

Refer back to the definition of general selfing **(Art. 2.2)**, its symbolic representation (2.5), and that of the split of the strengths of the individual mixes (2.6), and also **Fig. 2.3**.

Recollect the symbolic representation of two partially set mixes in selfing, namely,

$$A(R', T) = [(A(r_1, t_1))/(A(r_2, t_2))]_r \tag{3.25}$$

Selecting t_1 (refer to **Choice 1, Fig. 2.4**) as the time lag on which the final selfed strength is to be read, *i.e.*, $T = t_1$, $(t_1 > t_2)$, it requires only to express the resultant blend ratio R' in terms of those of the individual mixes r_1 and r_2, and that of the composite mix r.

Refer to (2.6) for split of the individual strengths $A(r_1, t_1)$ and $A(r_2, t_2)$.

Project the fresher strength $A(r_2, t_2)$ on time lag t_2, to get its equivalent strength $A(r_e, t_1)$, on time lag t_1, [where the blend ratio r_e (equivalent to r_2) is yet to be written in terms of r_2]. Then (3.25) when recast becomes

$$A(R', T) = [(A(r_1, t_1))/(A(r_e, t_1))]_r \tag{3.26}$$

where, r = blend ratio of the two partially set mixes $= \dfrac{(A(r_1, t_1))}{(A(r_e, t_1))} = \dfrac{Q_o}{Q_f}$

and, $Q_o = (A(r_1, t_1))$ = quantity of the old mix $= \dfrac{r}{1+r} Q$

$Q_f = (A(r_e, t_1))$ = quantity of the fresh mix $= \dfrac{1}{1+r} Q$

and, Q = total quantity of the composite $= Q_o + Q_f$ (3.27)

Now, in the individual mixes

$$A(r_1, t_1) = \text{strength of the old mix}$$

$$Q_o = (A(r_1, t_1)) = \text{quantity of the old mix}$$

This strength when viewed through pure selfing, yields

$$A(r_1, t_1) = [(A(\infty, t_1))/(A(0, t_1))]_{r1}\text{ , at time lag } t_1$$

and, $$r_1 = \text{blend ratio} = \frac{(A(\infty, t_1))}{(A(0, t_2))} = \frac{Q_{oo}}{Q_{of}}$$

where, $$Q_{o\,o} = (A(\infty, t_1)) = \text{older part of the old mix}$$

$$= \frac{r_1}{1+r_1} Q_o = \frac{r_1}{1+r_1} \frac{r}{1+r} Q$$

$$Q_{o\,f} = (A(0, t_1)) = \text{fresher part of the old mix}$$

$$= \frac{1}{1+r_1} Q_o = \frac{1}{1+r_1} \frac{r}{1+r} Q$$

and, $$\mathrm{Q}_o = Q_{oo} + Q_{o\,f} \tag{3.28}$$

$$A(r_e, t_1) = \text{strength of the fresh mix}$$

$$Q_f = (A(r_e, t_1)) = \text{(equivalent) quantity of the fresh mix}$$

This strength when viewed through pure selfing, yields

$$A(r_e, t_1) = [(A(\infty, t_1)) / (A(0, t_1))]_{re}\text{ , at time lag } t_1$$

and, $$r_e = \text{blend ratio} = \frac{(A(\infty, t_1))}{(A(0, t_1))} = \frac{Q_{fo}}{Q_{ff}}$$

where, $$Q_{f\,o} = (A(\infty, t_1)) = \text{older part of the fresh mix}$$

$$= \frac{r_e}{1+r_e} Q_f = \frac{r_e}{1+r_e} \frac{r}{1+r} Q$$

$$Q_{ff} = (A(0, t_1)) = \text{fresher part of the fresh mix}$$

$$= \frac{1}{1+r_e} Q_f = \frac{1}{1+r_e} \frac{r}{1+r} Q$$

and, $$Q_f = Q_{f\,o} + Q_{f\,f} \tag{3.29}$$

In the light of the Assumption-6, collect the older and fresher parts of the quantities from the blended mix (which is assumed to be a mechanical mixture of the ingredients) as given in (3.27)–(3.29), to get

$$R' = \frac{(Q)_{old}}{(Q)_{fresh}} = \frac{Q_{oo} + Q_{fo}}{Q_{of} + Q_{ff}} = \frac{\dfrac{r_1}{1+r_1}\dfrac{r}{1+r} + \dfrac{r_e}{1+r_e}\dfrac{1}{1+r}}{\dfrac{1}{1+r_1}\dfrac{r}{1+r} + \dfrac{1}{1+r_e}\dfrac{1}{1+r}}$$

$$\Rightarrow \qquad R' = \frac{r_e(1 + r_1 + rr_1) + rr_1}{1 + r + r_1 + rr_e} \qquad (3.30)$$

This is the required blend ratio for the resultant selfed mix when the strength is intended to be read at $t = t_1$, and is termed as **blend ratio equation**.

In **Fig. 3.9** the schematic diagram for R' for arriving at (3.30) is shown.

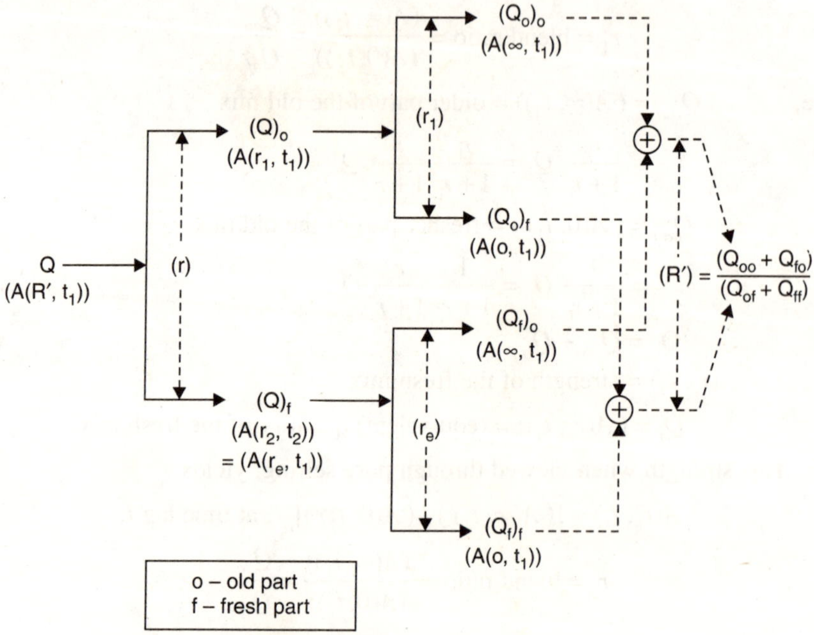

Fig. 3.9. Schematic diagram for resultant blend ratio R′

Similar relation of R' can be derived using other types of valid choices of the time lag as described in **Fig. 2.4**.

Case II: $r_1 \neq r_2$, $t_1 = t_2$, observed at $T = t_1$

Each one of the two partially set mixes are with different blend ratios r, but with same time lags t, and considered for selfing at the time lag t_1.

This case when symbolically written appear as

$$A(R', t_1) = [(A(r_1, t_1)) / (A(r_2, t_1))]_r$$

Recollect that, in Case I, the strength $A(r_2, t_2)$ is projected on t_1-ordinate to get equivalent strength $A(r_e, t_1)$ in terms of the new equivalent blend ratio r_e, (which is to be determined). Now in this case consider that, instead of $A(r_e, t_1)$ on t_1 for r_e, the given fresher strength is $A(r_2, t_1)$, and the other older strength viz., $A(r_1, t_1)$ remains in its same position. Thus substitute $r_e = r_2$ in (3.26), and following the remaining procedure, the required value for R' can be calculated. This can be directly obtained from (3.30) on substituting, $r_e = r_2$. Then

$$R' = \frac{r_2(1 + r_1 + rr_1) + rr_1}{1 + r + r_1 + rr_2} \tag{3.31}$$

Check for pure selfing case (from Case I and Case II)

In pure selfing, one of the mix is the OHC at time lag = 0, and other is the THC with blend ratio as ∞.

Consider OHC is the one having strength $A(r_2, t_2)$. Then put $t_2 = 0$, to get the strength of the OHC as $A(r_2, 0) = A(0, 0) = A(0, t_1)$, which turns out to be the strength of the IPC at $t = t_1$. Since $r_2 = 0$, consequently, $r_e = 0$.

The THC having strength $A(r_1, t_1)$ at $t = t_1$ and with $r_1 = \infty$, gives the strength as $A(\infty, t_1)$.

Substituting $r_e = 0$ in (3.27), or $r_2 = 0$ in (3.28), we get

$$R' = \frac{rr_1}{1 + r + r_1}$$

On dividing numerator and denominator by r_1

$$R' = \frac{r}{1 + \dfrac{1 + r}{r_1}}$$

and, further putting $r_1 = \infty$ (so that, $1/r_1 = 0$) in it, we get (at $t = t_1$) $\Rightarrow$ $R' = r$, which by definition, represents the blend ratio for pure selfing case (2.4).

Thus (3.25) reduces to

$$A(R', t_1) = [(A(\infty, t_1)) / (A(0, t_1))]_{R'}$$

$$\Rightarrow \quad A(r, t_1) = [(A(\infty, t_1)) / (A(0, t_1))]_r \qquad \textbf{(verified)}$$

Case III: $r_1 = \infty$; $(r_1 \neq r_2)$, $t_1 = t_2$, observed at $T = t_1$

One of the two partially set mixes is the basic one, while the other one is actually a blended one, but both the mixes are with the same time lag, and considered for selfing at the time lag t_1.

This case when symbolically written appears as

$$A(R', t_1) = [(A(\infty, t_1)) / (A(r_1, t_1))]_r$$

$$Q_o = (A(\infty, t_1)) = \text{quantity of the old mix} = Q_{o\,o} + Q_{of}$$

$$= [(A(\infty, t_1)) / (A(0, t_1))]_\infty$$

The blend ratio ∞, indicates that the fresh part in it is nil. That is,

$$Q_{of} = (A\ (0, t_1)) = 0.$$

Now $$Q_o = Q_{o\,o} + Q_{of} = \frac{r}{1 + r} Q + 0$$

$$Q_f = (A(0, t_1)) = \text{quantity of the fresh mix} = \frac{1}{1+r} Q$$

$$= Q_{fo} + Q_{ff} = \frac{1}{1+r} \cdot \frac{r_1}{1+r_1} Q + \frac{1}{1+r} \cdot \frac{1}{1+r_1} Q$$

$$R' = \frac{(Q)_{\text{old}}}{(Q)_{\text{fresh}}} = \frac{Q_{oo} + Q_{fo}}{Q_{of} + Q_{ff}} = \frac{\frac{1}{1+r} \times \left[r + \frac{r_1}{1+r_1} \right]}{\frac{1}{1+r} \times \left[\frac{1}{1+r_1} \right]}$$

$$= r + r_1 (1 + r)$$

$$\Rightarrow \qquad R' = r + r_1 (1 + r) \tag{3.32}$$

Check for pure selfing case (from Case III)

(3.32) when substituted by $r_1 = 0$, gives $R' = r$, which by definition is the criterion for pure selfing case (**verified**).

Graphical representation of the blend ratio equation:

The entire procedure of arriving at the expression of R', as given in (3.30), can be totally eliminated on adopting the following graphical procedure, which is simply based on the principle of similar triangles, usually employed for dividing a given straight line according to certain ratio, (say) m:n.

Construction: Refer to **Fig. 3.10 (*a*) – (*d*)**

Draw $AB = Q$.

To divide AB as AC: $CB = Q_o : Q_f = r$

Draw inclined line AX, and set off $AC_1 = m = r$, and $C_1B_1 = n = 1$, such that $AB_1 = m + n = 1 + r$. Join B_1B and draw C_1C parallel to it, to get point C on AB. Thus point C divides AB as $AC = Q_o$, and $CB = Q_f$.

Proof: We have, $r_1 = \dfrac{Q_o}{Q_f} \quad \Rightarrow \quad \dfrac{r_1}{1+r_1} = \dfrac{Q_o}{Q}$

From similar triangles ACC_1 and ABB_1, and with $AB = Q$,

$$\frac{AC}{AB} = \frac{AC_1}{AB_1} = \frac{m}{m+n} = \frac{r}{1+r} = \frac{Q_o}{Q}$$

$$\Rightarrow \qquad AC = Q_o \quad \text{and} \quad CB = AB - AC = 1 - Q_o = Q_f.$$

To divide AC as AD: $DC = Q_{oo}$: $Q_{of} = r_1$

Draw inclined line AY, and set off $AD_2 = a = r_1$, and $D_2C_2 = b = 1$, such that $AC_2 = a + b = 1 + r_1$. Join CC_2 and draw D_2D parallel to it, to get point D on AC. Thus point D divides AC as $AD = Q_{oo}$, and $DC = Q_{of}$.

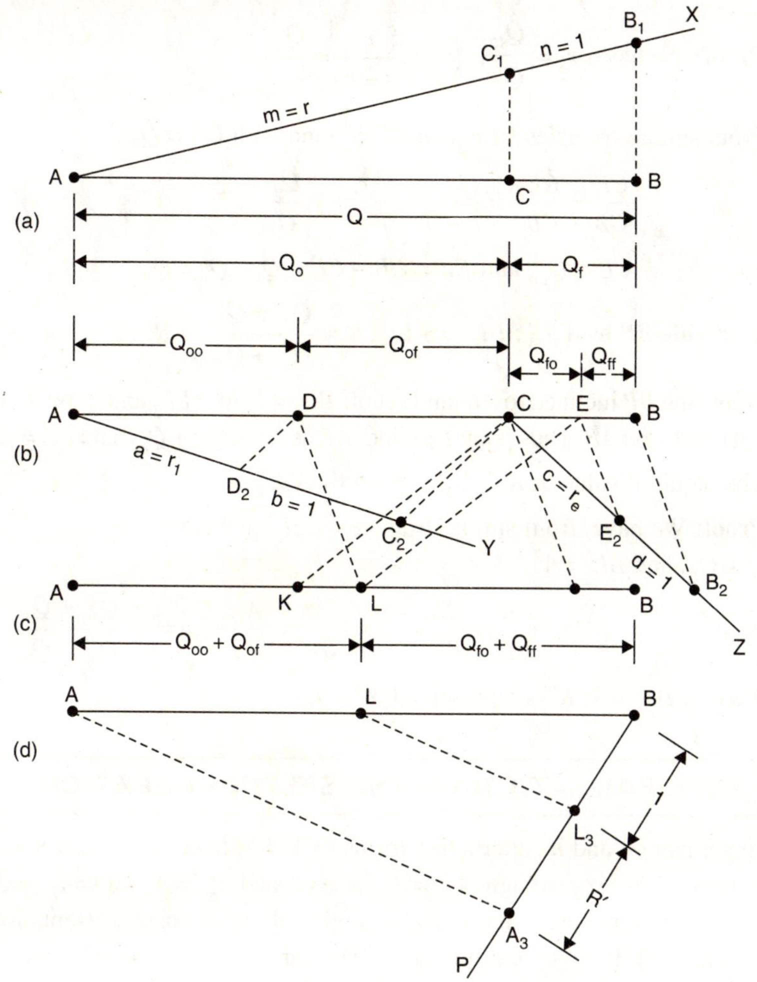

Fig. 3.10. Graphical representation of R'-equation

Proof: We have, $r_1 = \dfrac{Q_{oo}}{Q_{of}} \quad \Rightarrow \quad \dfrac{r_1}{1+r_1} = \dfrac{Q_{oo}}{Q_o}$

From similar triangles ADD_2 and ACC_2, and with $AC = Q_o$,

$$\frac{AD}{AC} = \frac{AD_2}{AC_2} = \frac{a}{a+b} = \frac{r_1}{1+r_1} = \frac{Q_{oo}}{Q_o}$$

$\Rightarrow \quad AD = Q_{oo}$, and $DC = AC - AD = Q_o - Q_{oo} = Q_{of}$.

To divide *CB* as *CE*: *EB* = Q_{fo}: $Q_{ff} = r_e$

Draw inclined line CZ, and set off $CE_2 = c = r_e$, and $E_2B_2 = d = 1$, such that $CB_2 = c + d = 1 + r_e$. Join BB_2 and draw E_2E parallel to it, to get point E on CB. Thus point E divides CB as $CE = Q_{fo}$, and $EB = Q_{ff}$.

Proof: We have, $r_e = \dfrac{Q_{fo}}{Q_{ff}} \quad \Rightarrow \quad \dfrac{r_e}{1+r_e} = \dfrac{Q_{fo}}{Q_f}$

From similar triangles CEE_2 and CBB_2, and with $CB = Q_f$,

$$\frac{CE}{CB} = \frac{CE_2}{CB_2} = \frac{c}{c+d} = \frac{r_e}{1+r_e} = \frac{Q_{fo}}{Q_f}$$

$$\Rightarrow \qquad CE = Q_{fo}, \text{ and } EB = CB - CE = Q_f - Q_{fo} = Q_{ff}.$$

To divide BP as $A_3L_3 : BL_3 = AL : LB = \dfrac{Q_{oo} + Q_{fo}}{Q_{of} + Q_{ff}} = R'$

Draw line BP inclined to AB, and set off $BL_3 = 1$. Join LL_3 and draw A_3A parallel to it, to get point L on AB. Thus point L divides AB as $AL = Q_{oo} + Q_{of}$, and $LB = Q_{fo} + Q_{ff}$.

The required value of R' is then given by A_3L_3.

Proof: We have, from similar triangles BLL_3 and BAA_3, $AL = Q_{oo} + Q_{of}$, and $LB = Q_{fo} + Q_{ff}$, and $BL_3 = 1$.

$$\Rightarrow \qquad \frac{AB}{BL} = \frac{BA_3}{BL_3} \quad \Rightarrow \quad \frac{AL}{BL} = \frac{A_3L_3}{BL_3} \quad \Rightarrow \quad \frac{Q_{oo} + Q_{fo}}{Q_{of} + Q_{ff}} = R'$$

Then, if $BL_3 = 1$, R' is represented by A_3L_3.

3.6 GENERALISATION OF THE SELFING EQUATION

The selfing equation and its alternative forms (3.4–3.6) have been derived using two boundary conditions viz., strength $A_{r,t} = A_{r,i}$ at $t = t_i$, and $A_{r,t} = A_{r,f}$ at $t = t_f$, and in view of this, the various representations namely, Graphical (or Vector) representation, Equal area representation, Rheological model (or Mechanical spring model representation) and Plate deflection analogy, have been proposed in **Art. 3.3**.

It can be seen that the same selfing equation (3.4) is equally valid when any two boundary conditions viz., strength $A_{r,t} = A_{r,tp}$ at $t = t_p$, and $A_{r,t} = A_{r,tq}$ at $t = t_q$, (where t_p and t_q are any two time lags in the time lag range, $t_i < t_p < t < t_q < t_f$), are chosen instead.

The geralisation is established first with the help of the **Vector form** of graphical construction (as described in **Fig. 3.2**), and then with the **$t - a_{rt}$ form** (as described in **Figs. 3.4, 3.4*A* and 3.4*B***), through items (*a*) and (*b*) respectively as follows:

(*a*) Using vector form of graphical construction:

The graphical construction of the selfing equation is described in **Fig. 3.2** in **Art. 3.3**, where it is intended to obtain $A_{r,t}$ (GT_2) at t from the given $A_{r,i}$ (AI_1) at t_i and

$A_{r,f}$ ($AF_1 = EF_2$) at t_f. The rays through the pole O viz., OF, OI and OT identifying the reciprocal of the strengths $a_{r,f}$, $a_{r,i}$ and $A_{r,t}$, respectively, on ML (axis through t_i), are drawn corresponding to the closed vector triangle ABC as, $AB + BC = AC$. In this case it is to note, that the two extreme strengths are $A_{r,i}$ and $A_{r,f}$ at the two extreme setting times t_i and t_f, and t is in between these two setting times. All other representations of the selfing equation presented earlier, use these two boundary values.

It can be easily shown that, the said graphical construction is equally valid for getting the strength ordinate $A_{r,t}$ using the strengths $A_{r,p}$ and $A_{r,q}$ at time lags $t = t_p$ and $t = t_q$, where, $t_i < t_p < t < t_q < t_f$. In the following, five graphical constructions are discussed, through **Constructions 1–5**.

Construction: Refer to **Fig. 3.11**.

Construction 1: To obtain A_{r,t_2} at t_2; from $A_{r,i}$ at t_i, and $A_{r,f}$ at t_f; ($t_i < t_2 < t_f$)

Given $AM = A_{r,i}$ at t_i, $ET_f = AN = A_{r,f}$ at t_f. To get $G_2T_2 = A_{r,t_2}$ at t_2.

Follow the same procedure as done for **Fig. 3.2**.

Locate $AM_1 = a_{r,i} = 1/AM$, $AN_1 = a_{r,f} = 1/AN$.

Draw vector triangle AH_1L_1, where AH_1 and H_1L_1 are parallels to ON_1 (at t_2) and OM_1 (at t_f), respectively,

so that vectorially,

$$AH_1 + H_1L_1 = AL_1.$$

Draw OQ_1 parallel to the closing line AL_1, and then draw OQ perpendicular to OQ_1 to get $AQ = G_2T_2 = A_{r,t_2}$ at t_2.

Construction 2: To obtain A_{r,t_1} at t_1 ; from $A_{r,i}$ at t_i , and $A_{r,f}$ at t_f; ($t_i < t_1 < t_f$)

Given $AM = A_{r,i}$ at t_i, $ET_f = AN = A_{r,f}$ at t_f. To get $G_1T_1 = A_{r,t_1}$ at t_1.

Follow the same procedure as done for **Fig. 3.2** or for **Construction 1.**

Locate $AM_1 = a_{r,i} = 1/AM$, $AN_1 = a_{r,f} = 1/AN$.

Draw vector triangle AH_2L_2 , where AH_2 and H_2L_2 are parallels to ON_1 (at t_1) and OM_1 (at t_f), respectively,

so that vectorially, $$AH_2 + H_2L_2 = AL_2.$$

Draw OP_1 parallel to the closing line AL_2, and then draw OP perpendicular to OP_1 to get $AP = G_1T_1 = A_{r,t_1}$ at t_1.

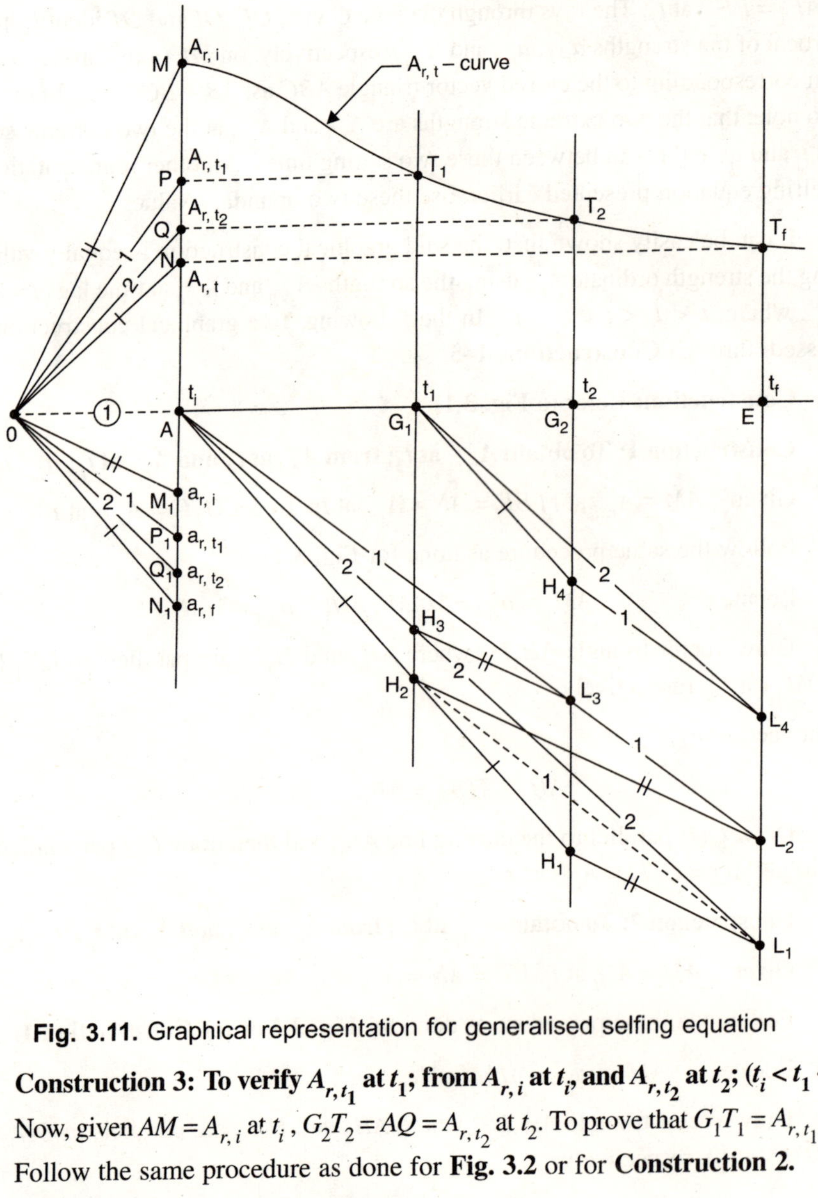

Fig. 3.11. Graphical representation for generalised selfing equation

Construction 3: To verify A_{r,t_1} at t_1; from $A_{r,i}$ at t_i, and A_{r,t_2} at t_2; $(t_i < t_1 < t_2)$

Now, given $AM = A_{r,i}$ at t_i, $G_2T_2 = AQ = A_{r,t_2}$ at t_2. To prove that $G_1T_1 = A_{r,t_1}$ at t_1.

Follow the same procedure as done for **Fig. 3.2** or for **Construction 2.**

Read the strength at t_2 as if that at t_f.

Locate $AM_1 = a_{r,i} = 1/AM$, $AQ_1 = a_{r,t_2} = 1/AQ$.

Draw vector triangle AH_3L_3 , where AH_3 and H_3L_3 are parallels to OQ_1 (at t_1) and OM_1 (at t_2), respectively,

so that vectorially, $$AH_3 + H_3L_3 = AL_3.$$

The closing line is AL_3, (which is a line parallel to OP_1). Then draw OP as perpendicular to OP_1. Hence the same strength $AP = G_1T_1 = A_{r,\,t_1}$ at t_1 results.

Construction 4: To verify $A_{r,\,t_2}$ at t_2; from A_{r,t_1} at t_1, and $A_{r,\,f}$ at t_f; ($t_1 < t_2 < t_f$)

Now, given $G_1T_1 = AP = A_{r,\,t_1}$ at t_1, $ET_f = AN = A_{r,f}$ at t_f. To prove that $G_2T_2 = A_{r,\,t_2}$ at t_2.

Follow the same procedure as done for **Fig. 3.2** or for **Construction 3.**

Read that the strength at t_1 as if that at t_i.

Locate $AP_1 = a_{r,\,t_1} = 1/AP, AN_1 = a_{r,f} = 1/AN.$

Draw vector triangle $G_1H_4L_4$, where G_1H_4 is parallel to ON_1 (at t_f), and so also to AH_2, and H_4L_4 is parallel OP_1 (at t_1), respectively,

so that vectorially, in triangle $G_1H_4L_4$, $G_1H_4 + H_4L_4 = G_1L_4$.

Compare two vector triangles AH_2L_2 and $H_2H_1L_1$ whose corresponding sides AH_2 and H_2L_2 are parallel to H_2H_1 and H_1L_1, respectively.

Thus: $AH_2 + H_2L_2 = AL_2 \Rightarrow H_2H_1 + H_1L_1 = H_2L_1 \Rightarrow H_2L_1$ is parallel to AL_2.

Now compare two similar vector triangles $G_1H_4L_4$ and AH_2L_1 whose three corresponding sides are parallel to one another.

Then,
$$AH_2 + H_2L_1 = AL_1 \Rightarrow H_2H_1 + H_1L_1 = H_2L_1$$
$$\Rightarrow G_1H_4 + H_4L_4 = G_1L_4.$$

This confirms that the closing line G_1L_4 is parallel to OQ_1, and thus the resultant strength $AQ = G_2T_2 = A_{r,\,t_2}$ at t_2 is verified.

Construction 5: (not shown) **To verify $A_{r,t}$ at t; from A_{r,t_1} at t_1, and A_{r,t_2} at t_2; ($t_1 < t < t_2$)**

Now, given $G_1T_1 = AP = A_{r,\,t_1}$ at t_1, $G_2T_2 = AQ = A_{r,\,t_2}$ at t_2. To get, $A_{r,\,t}$ at t, where t lies in between t_1 and t_2.

Follow the same procedure as done for **Fig. 3.2** or for **Construction 1** on replacing t_i and t_f by t_1 and t_2, respectively.

However, this is not necessary. **Construction 1** to **Construction 4** carry sufficient proof for this case too.

(*b*) Using t – a_{rt} form of graphical construction:

Refer to **Art. 3.3** and **Fig. 3.4** and recollect the inference that, for any strength of certain value of r and observed at time t, the junction point of the $t - a_{rt}$ pair must lie on the diagonal of the $t - a_{rt}$ rectangle.

Applying this statement for Curve 1 of **Fig. 3.4*B***, for example, note that:

CE is the diagonal of the $t - a_{rt}$ rectangle $CDEF$, and the boundary t – lines are CF (at $t = t_i$) and DE (at $t = t_f$).

If for the sake of convenience, the two t – lines at t_3 and t_1 (corresponding to strength locations at S and P), are renamed as t_p and t_q, respectively, the $t - a_{rt}$ rectangle will have the diagonal S_1P_1 which will pass through S_1 and P_1 and lie on CE. Then the strength at t_2 corresponding to R on Curve 1, if requited to be observed, will be at R_1 on the same diagonal $S_1R_1P_1$, which in other words, is the same line CE.

This observation establishes the generalization of the selfing equation.

Remarks: Following **Fig. 3.11** or **Fig. 3.4*B***, it can be readily noticed that:

For any strength-time curve with a certain r-value, the graphical construction as suggested in Fig. 3.2 or Fig. 3.4 for obtaining the strength at any time lag t, lying between any two time lags t_p and t_q using the corresponding strengths at these two instants, where $t_i < t_p < t < t_q < t_f$, is always valid.

In consequence to this, the A_{rt} -equation and its alternative forms (3.4–3.6) can be **recast** as:

$$A_{r,t} = \frac{A_{r,tp} \cdot A_{r,tq} \cdot (t_q - t_p)}{t \cdot (A_{r,tp} - A_{r,tq}) + A_{r,tq} \cdot t_q - A_{r,tp} \cdot t_p} \quad (3.33)$$

or,

$$\frac{t_q - t_p}{A_{r,t}} = \frac{t - t_p}{A_{r,tq}} + \frac{t_q - t}{A_{r,tp}} \quad (3.34)$$

and,

$$(t_q - t) \cdot \left[\frac{1}{A_{r,t}} - \frac{1}{A_{r,tp}}\right] = (t - t_p) \cdot \left[\frac{1}{A_{r,tq}} - \frac{1}{A_{r,t}}\right]$$

i.e.,

$$(t_q - t) \cdot (a_{r,t} - a_{r,tp}) = (t - t_p) \cdot (a_{r,tq} - a_{r,t}) \quad (3.35)$$

Thus **the Graphical (Vector, or, $t - a_{rt}$) representation, Equal area representation, Rheological model (or mechanical spring model representation) and Plate deflection analogy as proposed in Art. 3.3, can well be translated as the generalised form through (3.33–3.35) on replacing simply t_i and t_f in (3.4–3.6) by t_p and t_q, respectively.**

CHAPTER 4

Analytical Solution of Selfing and Crossing Problems

4.1 GENERAL

The calculation procedure of the various stages of Selfing as well as Crossing of two partially set mixes in blending requires **four** items viz., **setting times, characteristic strength parameters** (*i.e.*, **CSP-s), Selfing equation** and **blend ratio equation.** They are:

for each mix:

1. **Setting times:** t_i and t_f.
2. **Characteristic strength parameters** (**CSP-s**): slopes m_k-s and intercepts C_k-s of the linear form of the **ArkR-plot** (*i.e.*, variation of the A_{rk} vs. R, as given by (3.20) or (3.21), namely

$$A_{rk} = C_k + m_k R \tag{4.1}$$

i.e.,

$$A_{ri} = C_i + m_i .R \text{ , and, } A_{rf} = C_f + m_f .R \tag{4.2}$$

(where, the suffix k stands for i or f, representing time lags t_i, and t_f, respectively; and R is any suitable function of the blend ratio r).

This relation will be referred to as **ArkR-equation.**

formulae to use:

3. **Selfing equation:** (3.4) or (3.5) or (3.6)

$$A_{r,t} = \frac{A_{r,i} \cdot A_{r,f} \cdot (t_f - t_i)}{t.(A_{r,i} - A_{r,f}) + A_{r,f} \cdot t_f - A_{r.i} \cdot t_i}$$

or, $$A_{r,t} = \frac{A_{r,i} \, .A_{r,f} \, .(t_f - t_i)}{A_{r,i} \, (t - t_i) + A_{r,f} \, (t_f - t)} \quad (4.3)$$

which, for two Zones $(0 - t_i)$ and $(t_i - t_f)$ can be written as

for **Zone I** : $$A_{r,t} = \frac{A_{r,0} \, .A_{r,i} \, .(t_i)}{t.A_{r,0} + A_{r,i} \, .(t_i - t)}$$

for **Zone II** : $$A_{r,t} = \frac{A_{r,i} \, .A_{r,f} \, .(t_f - t_i)}{t.(A_{r,i} - A_{r,f}) + A_{r,f} \, .t_f - A_{r.i} \, .t_i}$$

or $$\frac{t_f - t_i}{A_{r,t}} = \frac{t - t_i}{A_{r,f}} + \frac{t_f - t}{A_{r,i}} \quad (4.4)$$

or $$(t_f - t) . \left[\frac{1}{A_{r,t}} - \frac{1}{A_{r,i}} \right] = (t - t_i) . \left[\frac{1}{A_{r,f}} - \frac{1}{A_{r,t}} \right] \quad (4.5)$$

4. **Blend ratio equation:** (3.30)

$$R' = \frac{r_e \, (1 + r_1 + r r_1)}{1 + r + r_1 + r_e}, \text{ (choice of the time lag is at } t = t_1) \quad (4.6)$$

(or similar suitable one, depending on the choice of the time lag).

4.2 SELFING IN A SINGLE MIX

[Note : Strength, if not specifically mentioned, will indicate compressive strength, f_c]

Example 4.1. *A sand-cement mortar shows Type I variation of strength with time lag. Given that, $t_i = 1$ hr, $t_f = 4$ hr (**Fig. 4.1**). The CSP-s (in kg/sq cm) viz., (C_i, m_i) and (C_f, m_f) are (180.62, 29.61) and (135.43, 43.10). In the ArkR-plot, viz., $A_{rk} = C_k + m_k R$, R is taken as 1/r.*

***Determine** the strength for the basic partially set mix at t = 3 hr, and percentage reduction in strength with respect to that at t_i.*

Solution. The problem requires to find any strength at t in between t_i and t_f, when the two strengths at t_i and t_f are given, for the same curve.

Here, it is for the basic partially set mix which means that, at any time, no fresh mix is blended to it. That is, its $r = \infty$.

Given: The two strength values at t_i and t_f for the basic mix ($r = \infty$), calculate the strength at $t = 3$, for the same curve $r = \infty$.

Rewrite the Selfing equation (4.3), with $r = R'$ in the LHS.

$$A(R', t) = \frac{A_{r,i} \cdot A_{r,f} \cdot (t_f - t_i)}{t.(A_{r,i} - A_{r,f}) + A_{r,f}.t_f - A_{r,i} .t_i}.$$

To find: The final strength $A(R', t)$, substitute $r = R' = \infty$ and $t = 3$, to the LHS of the equation to get $A(\infty, 3)$.

Since, $t = 3$ lies between the given t_i and t_f values, the Selfing formula is applicable to Zone II only.

Since, for $r = \infty$, $1/r = 0$, the two intercepts $C_k - s$ are the strengths $A(\infty, t_i)$ and $A(\infty, t_f)$ of the very basic mix.

So, $\quad C_i = A(\infty, t_i) = 180.62, \quad$ and $\quad C_f = A(\infty, t_f) = 135.43$

Now, $\quad A_{ri} = C_i + m_i .R = C_i + m_i .\frac{1}{r} = 180.62 + \frac{29.61}{r}$

and, $\quad A_{rf} = C_f + m_f .R = C_f + m_f .\frac{1}{r} = 135.43 + \frac{43.10}{r}$

for $\quad r = \infty, A_{r,i} = 180.62, \quad$ and $\quad A_{r,i} = 135.43.$

Substitute these along with $t = 3$ in the RHS of the equation

$$A(\infty, 3) = \frac{180.62 \times 135.43 \times (4-1)}{3 \times (180.62 - 135.43) + 4 \times 135.43 - 1 \times 180.62} = \mathbf{123}$$

Now, the **reduction in strength** $= \dfrac{180.62 - 123}{180.62} \times 100 = \mathbf{31.9}\ \%$

The required strength of the basic partially set mix at $t = 3$ hr is, 123 kg/sq cm, which has reduced by 31.9%, from that at the initial setting time, on lying exposed to the open atmosphere for two more hours after setting began, before casting.

Example 4.2. *Determine the strength of the partially set mix at t = 3 hr for r = 2. The other data is same as in Example 4.1.*

A mix (same data as in Example 4.1) weighing 80 kg, shows reduction in strength after lying exposed up to 3 hr before casting. At that time lag, 40 kg of the same mix type was instantly prepared, thoroughly blended to the said partially set mix, and was cast.

Determine the strength expected from the selfed mix, and the gain in strength in comparison to that of the unblended one.

Solution. The problem requires to find the strength at any time lag t on a different blend ratio curve, when only the two strengths at t_i and t_f, of the basic mix, are given.

Refer to **Fig. 4.1.**

In Example 4.1, we calculated $A(\infty, 3)$, when $A(\infty, t_i)$ and $A(\infty, t_f)$ were given for the curve $r = \infty$. Here, we shall calculate $A(2, 3)$, when $A(2, t_i)$ and $A(2, t_f)$ are given for the curve $r = 2$.

For $r = 2$, the ArkR-equation yields

$$A(2, t_i) = C_i + m_i \cdot \frac{1}{r} = 180.62 + \frac{29.61}{2} = 195.42$$

$$A(2, t_f) = C_f + m_f \cdot \frac{1}{r} = 135.43 + \frac{43.10}{2} = 156.98$$

Recall the Selfing equation (4.3). Substitute $t_i = 1$ hr, $t_f = 4$ hr, $t = 3$ hr, and $r = 2$ in the RHS, and the LHS is the strength $A(2, 3)$ to be determined. Thus

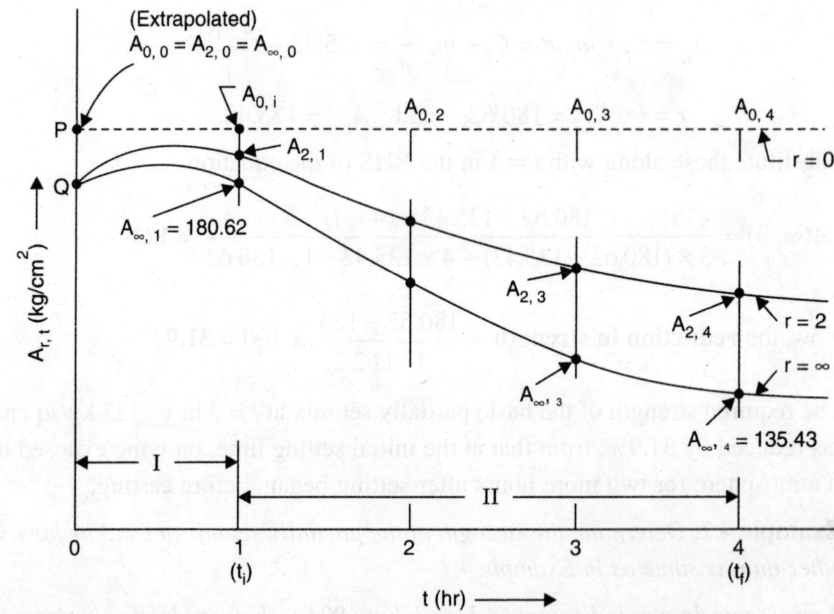

Fig. 4.1. Strength in terms of those at t_i and t_f

$$A(2, 3) = \frac{A(2, t_i) \times A(2, t_f) \times (4 - 1)}{(3) \times [A(2, t_i) - A(2, t_f)] + (4) \times A(2, t_f) - (1) \times A(2, t_i)}$$

$$\Rightarrow \quad A(2, 3) = \frac{(195.42) \times (156.98) \times (3)}{(3) \times [195.42 - 156.98] + (4) \times (156.98) - (1) \times (195.42)}$$

$$= \mathbf{167.99}$$

The **required strength** of the partially set mix at $t = 3$, for $r = 2$ is **167.99** kg/sq cm.

Now, the **enhancement in strength** $= \dfrac{167.99 - 123}{123} \times 100 =$ **36.59** %, compared to $A(\infty, 3)$.

Still, the **reduction in strength** $= \dfrac{180.62 - 167.99}{180.62} \times 100 =$ **6.98** %, compared to that at t_i.

Example 4.3. *A composite mix is formed out of blending of some quantity of the partially set mix at t = 3 hr with some instantaneously prepared fresh mix of the same mix type, at that instant of time lag, in the proportion of 2 : 1 by weight. Determine the IPC strength A(0, 3), if the strength A(2, 3) at t = 3 hr is 167.99 kg/sq cm. The other data is same as in Example 4.2.*

Solution. The problem requires to find the IPC strength $A(0, 3)$, if the THC strength $A(2, 3)$ is given with the knowledge of that of the basic strength $A(\infty, 3)$ and the blend ratio. (In this sense, it is the reverse problem to Example 4.2).

Refer to **Fig. 4.1.**

Here, the blend ratio r = (wt. of old quantity) / (wt. of fresh quantity)

$$= (A(\infty, 3)) / (A(0, 3)) = 2$$

Strength and quantity-wise, this problem can be expressed through Selfing as

$$A(2, 3) = [\ (A(\infty, 3)) / (A(0, 3))\ ]_2$$

in which, $A(2, 3)$ in the LHS = 167.99 kg/sq cm, and $A(\infty, 3)$ in the RHS = 123. kg/sq cm.

It is required to **find** $A(0, 3)$.

Note: Relation $A(0, 0) = A(\infty, 0) = A(2, 0) = A(r, 0)$ of OHC, which is also = $A(0, 3)$ of IPC, is only valid for the ideal variation of the strength vs. time lag, namely, for Type I where, the horizontal line for $r = 0$, through point P (**Fig. 4.1**), representing $A(0, 0)$, occupies the top most level, so that, all other strengths are less than this value.

For other types of variation, it is not expected, and in most cases, often $A(0, 0) < A(r, t_i)$, as represented by point Q in **Fig. 4.1**. Then, in such cases, $A(0, 0)$ value must be known along with CSP-s.

However, whether given or not, $A(0, 0)$ can be approximately located in Zone I, by extrapolating the corresponding strength variations $A(r, t_i)$ at t_i, on the ordinate at $t = 0$ (point P in **Fig. 4.1**), and this hypothetical value can be used in order to satisfy the equality of the strengths of IPC and OHC at the location t, in Zone II.

Since, information of the strength at $t = 0$, (which often is the strength of the fresh mix) is not available in this problem, we take, $A(0, 0) = A(\infty, 0) = A(2, 0) = A(0, 3)$. This, in other words, means that Zone II as the portion of Type I variation, in the entire time lag range, $t = 0$ to $t = t_f$.

So substitute, $A(r, t_i) = A(0, 3) = A(2, 0) = A(0, 0) \rightarrow$ **to be determined**

$$A(r, t_f) = A(2, 4) = 156.98 \rightarrow \textbf{given}$$

$$t_i = 0,\ t_f = 4,\ t = 3, \text{ and } r = 2 \rightarrow \textbf{given}$$

in the Selfing equation (4. 3),

$$A(2, 3) = \frac{A(2,0) \times A(2,4) \times (4-0)}{(3) \times [A(2,0) - A(2,4)] + (4) \times A(2,4) - (0) \times A(0,0)}$$

$$\Rightarrow \quad 167.99 = \frac{A(0,0) \times (156.98) \times (4)}{(3) \times [A(0,0) - (156.98)] + (4) \times (156.98)}$$

$$\Rightarrow \quad 123.95\ A_{00} = 26371.07$$

$$\Rightarrow \quad A_{00} = A(0, 3) = \mathbf{212.76}$$

The **required strength A(0, 3) = 212.76** kg/sq cm.

This means that, (W) kg of an old partially set mix (showing strength of 123. kg/sq cm), at $t = 3$ hr of open exposure before casting, needs blending of (W / 2) kg of fresh mix instantly prepared at the same instant of time (showing strength of 212.76 kg/sq cm) to it, to yield a strength of 167.99 kg/sq cm.

The **strength enhancement** $= \dfrac{167.99 - 123}{123} \times 100\% = \mathbf{36.5}\ \%$.

Example 4.4. *A 1:3 mortar mix with 0.5 w/c has the CSP-s (in kg/sq cm):*

$$C_i = 238,\ m_i = -18.33,\ C_f = 211,\ m_f = 102.$$

The two setting times are: $t_i = 1$ *hr,* $t_f = 3$ *hr.*

In the ArkR-plot, R is taken as $\log_{10}(1/r)$.

Determine the strengths for the mix A(4, 2.5) and A(2, 1.5).

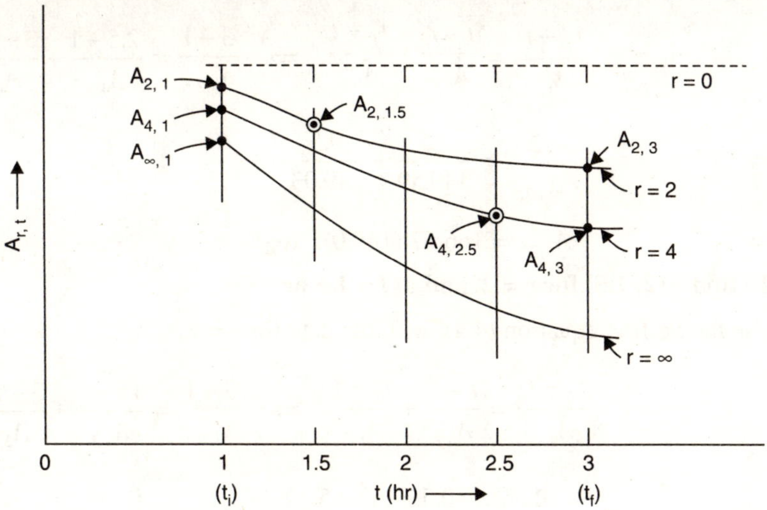

Fig. 4.2. Selfed strength at given *r* and *t*

Solution.

Refer to **Fig. 4.2.**

We have the ArkR-equation

$$A_{r,i} = C_i + m_i \log_{10} \frac{1}{r}$$

$$A_{r,f} = C_f + m_f \log_{10} \frac{1}{r}$$

For $r = 4$, $t_f = 3$ hr, the ArkR-equation yields

$$A_{4,1} = 238 - 18.33 \log_{10} \frac{1}{4} = 249.03$$

$$A_{4,3} = 211 + 102 \log_{10} \frac{1}{4} = 149.59$$

For $r = 2$, $t_f = 3$ hr, the ArkR-equation yields

$$A_{2,1} = 238 - 18.33 \log_{10} \frac{1}{2} = 243.52$$

$$A_{2,3} = 211 + 102 \log_{10} \frac{1}{2} = 180.29.$$

To find $A(4, 2.5)$ for $r = 4$, and at $t = 2.5$ hr:

Use the Selfing equation (4.4), with the data for $r = 4$, $t_f = 3$

$$\frac{t_f - t_i}{A_{r,t}} = \frac{t - t_i}{A_{r,f}} + \frac{t_f - t}{A_{r,i}} \quad \Rightarrow \quad \frac{3-1}{A_{4,2.5}} = \frac{2.5-1}{A_{4,3}} + \frac{3-2.5}{A_{4,1}}$$

$$\Rightarrow \quad \frac{2}{A_{4,2.5}} = \frac{1.5}{149.59} + \frac{0.5}{249.03}$$

$$\Rightarrow \quad A_{4,2.5} = 166.67\ (166.0).$$

To find $A(2, 1.5)$ for $r = 2$, and at $t = 1.5$ hr:

Use the Selfing equation (4.4), with the data for $r = 2$, $t_f = 3$

$$\frac{t_f - t_i}{A_{r,t}} = \frac{t - t_i}{A_{r,f}} + \frac{t_f - t}{A_{r,i}} \quad \Rightarrow \quad \frac{3-1}{A_{2,2.5}} = \frac{1.5-1}{A_{2,3}} + \frac{3-1.5}{A_{2,1}}$$

$$\Rightarrow \quad \frac{2}{A_{2,1.5}} = \frac{0.5}{180.29} + \frac{1.5}{243.52}$$

$$\Rightarrow \quad A_{2,\,1.5} = \mathbf{223.9\ (219.0)}$$

Alternatively, using (4.3)

$$A_{2,\,1.5} = \frac{(243.52) \times (180.29) \times (3-1)}{(1.5) \times (243.52 - 180.29) + (3) \times (180.29) - (1) \times (234.52)}$$

$$= \frac{87808.44}{392.195} = \mathbf{223.89}$$

[The values within ( ) are the **experimental** ones].

Example 4.5. *A mortar mix 1: 4 was prepared with w/c ratio of 0.6. The two setting times (in hr) are 1.75 and 3. The CSP-s (in kg/sq cm) viz., (C_i, m_i) and (C_f, m_f) are (160, 15) and (105, 35), respectively. In the ArkR-plot, viz., $A_{rk} = C_k + m_k R$, where, R is taken as 1/r.*

(*a*) ***Calculate*** *the strength of the basic mix after exposure of 2 hr in open atmosphere before casting.*

(*b*) ***Calculate*** *the strength of the basic mix after exposure of 2.67 hr in open atmosphere before casting.*

(*c*) *If the basic mix at time lag 2.67 hr is blended with some freshly prepared mix of the same mix type, in order to regain the same strength as that of the basic mix at time lag of 2 hr,* ***determine*** *the proportion of the IPC to that of the THC.*

Solution. $A_{ri} = C_i + m_i .R = C_i + m_i .\frac{1}{r}$

$$A_{rf} = C_f + m_f .R = C_f + m_f .\frac{1}{r}$$

This problem requires to find the equivalent blend ratio r_e needed so that the strength $A(r_e, 2.67)$ at $t = 2.67$ hr, will be same as that of the basic mix at $t = 2$ hr *i.e.*, of $A(\infty, 2)$.

Refer to **Fig. 4.3.**

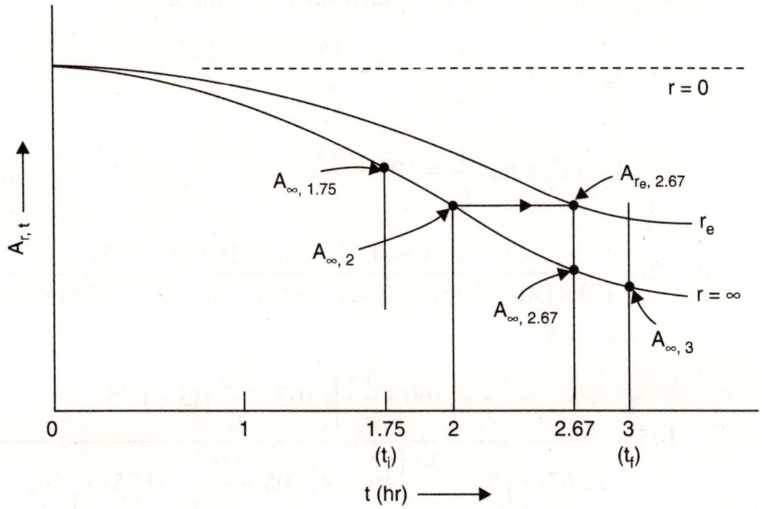

Fig. 4.3. Equivalent blend ratio corresponding to a strength

We need to **find** $A(\infty, 2)$ as well as $A(r_e, 2.67)$ expressed in terms of the unknown r_e, and get the required r_e on equating these two strengths.

Here, $r = \infty$, (*i.e.*, $1/r = 0$), refers to the basic mix.

So, $C_i = A(\infty, t_i) = 160$, and $C_f = A(\infty, t_f) = 105$.

(*a*) **To find $A(\infty, 2)$ for $r = \infty$, and at $t = 2$ hr:**

Using Selfing equation (4.3) with $t_i = 1.75$, $t_f = 3$, $r = \infty$, $t = 2$

$$A(\infty, 2) = \frac{A(\infty, 1.75) \times A(\infty, 3) \times (3 - 1.75)}{(2) \times [A(\infty, 1.75) - A(\infty, 3)] + (3) \times A(\infty, 3) - (1.75) \times A(\infty, 1.75)}$$

$$\Rightarrow \quad A(\infty, 2) = \frac{(160) \times (105) \times (1.25)}{(2) \times [(160) - (105)] + (3) \times (105) - (1.75) \times (160)}$$

$$= 145 \text{ kg/sq cm.}$$

(*b*) **To find $A(\infty, 2.67)$ for $r = \infty$, and at $t = 2.67$ hr:**

Similarly using Selfing equation (4.3) with $t_i = 1.75$, $t_f = 3$, $r = \infty$, $t = 2.67$

$$A(\infty, 2.67) = \frac{(160) \times (105) \times (1.25)}{(2.67) \times [(160) - (105)] + (3) \times (105) - (1.75) \times (160)}$$

$$= 111.5 \text{ kg/sq cm.}$$

[**Note**: The use of **differential** to approximate the value of $A(\infty, 2.67)$ at t = 2.67, from $A(\infty, 2)$ at t = 2, has been shown latter in **Example 4.19**]

(*c*) **To find r_e, when $A(r_e, 2.67) = A(\infty, 2)$:**

Refer (4.3) with LHS as $A(\infty, 2) = 145$,

and in RHS, with r_e as unknown **to be determined**, substitute

$$A_{ri} = C_i + m_i \cdot \frac{1}{r_e} = 160 + \frac{15}{r_e}$$

$$A_{rf} = C_f + m_f \cdot \frac{1}{r_e} = 105 + \frac{35}{r_e}$$

$$\Rightarrow \quad A(\infty, 2) = \frac{A(\infty, 1.75) \times A(\infty, 3) \times (3 - 1.75)}{(2) \times [A(\infty, 1.75) - A(\infty, 3)] + (3) \times A(\infty, 3) - (1.75) \times A(\infty, 1.75)}$$

$$\Rightarrow \qquad 145 = \frac{\left(160 + \frac{15}{r_e}\right)\left(105 + \frac{35}{r_e}\right)(3 - 1.75)}{(2.67) \times \left[55 - \frac{20}{r_e}\right] + (3) \times \left(105 + \frac{35}{r_e}\right) - (1.75) \times \left(160 + \frac{15}{r_e}\right)}.$$

To **solve** for r_e, let us rewrite this quadratic equation after proper simplification as

$$a\left(\frac{1}{r_e}\right)^2 + b.\left(\frac{1}{r_e}\right) + c = 0$$

where, *a*, *b* and *c* are coefficients given by

$$a = m_i \times m_f = 15 \times 35 = 525$$

$$b = A(\infty, t_i).m_f + A(\infty, t_f).m_i - \frac{A(r, t)}{t_f - t_i}.[(t - t_i).m_i + (t_f - t).m_f]$$

$$= 160 \times 35 + 105 \times 15 - \frac{145}{3 - 1.75}[0.92 \times 15 + 0.33 \times 35]$$

$$= 4234.4$$

$$c = A(\infty, t_i).\,A(\infty, t_f) - \frac{A(r, t)}{t_f - t_i}.[(t - t_i)\,.\,A(\infty, t_i) + (t_f - t).\,A(\infty, t_f)]$$

$$= 160 \times 105 - \frac{145}{3 - 1.75}[0.92 \times 160 + 0.33 \times 105]$$

$$= -4294.6$$

Thus solving for $x = \frac{1}{r_e}$, from

$$525\,x^2 + 4234.4\,x - 4294.6 = 0$$

$$x = \frac{-4234.4 \pm 5191.22}{1050} = 0.911 = \frac{1}{r_e}\text{, (take the + ve value only)}$$

$\Rightarrow$ $r_e = 1.098 = \mathbf{1.1}$ (say).

The **required equivalent blend ratio, r_e = 1.1** for the required Selfed strength at $t = 2.67$ hr.

Check: To verify whether A(1.1, 2.67) is equivalent to $A(\infty$, 2) or not:

Substitute, $t = 2.67$, $r = r_e = 1.1$, and

$$A_{ri} = 160 + \frac{15}{r_e} = 160 + \frac{15}{1.1} = 173.64$$

$$A_{rf} = 105 + \frac{35}{r_e} = 105 + \frac{35}{1.1} = 136.82$$

in the Selfing equation (4.3), we get

$$A_{1.1,\,2.67} = \frac{173.64 \times 136.82 \times (3 - 1.75)}{2.67 \times (173.64 - 136.82) + 3 \times 136.82 - 1.75 \times 173.64}$$

$$= \frac{29696.781}{98.31 + 410.46 - 303.87} = \frac{29696.781}{204.9} = \mathbf{144.933}\text{ kg/sq cm.}$$

This value **exactly tallies** with that of $A(\infty$, **2) = 145** kg/sq cm. (**Checked**).

Example 4.6. *Standard samples were prepared for conducting pull-out bond test by inserting 10 mm diameter mild steel rod in concrete block of M 20 grade. The curing was performed for the samples by allowing first 14 days in water, and balance of the next 14 days in air, instead of full 28 days of water curing. The strength vs time lag variations of the bond strengths (f_b, in kg/sq cm), for various blend ratios r, and time lags t, were found to conform Type II variation.*

In the ArkR-plot, viz., $A_{rk} = C_k + m_k R$, where, R is taken as $\log_{10}(r)$.

The other data are:

$$A_{00} = 79.58,\ C_i = 97.20,\ m_i = -8.293,\ C_f = 69.20,\ m_f = -6.282,$$

and the two setting times are, t_i = 1 hr, t_f = 3 hr.

***Determine**: For what blend ratio r, the bond strength should show 29.15 kg/sq cm at a time lag of 2.167 hr?*

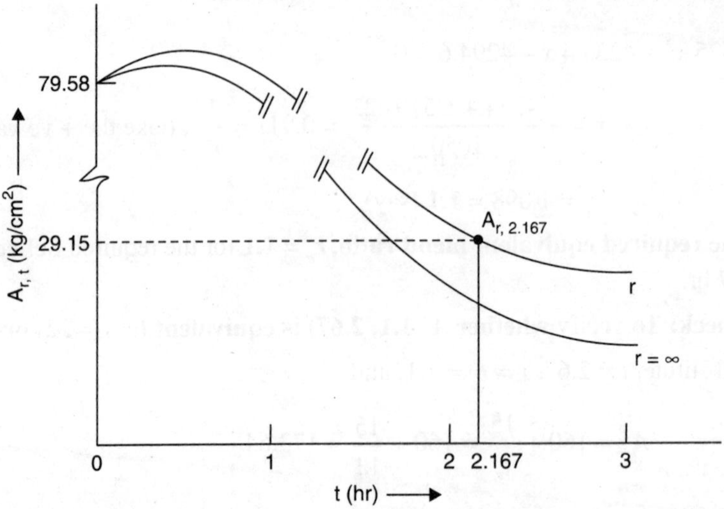

Fig. 4.4. Blend ratio corresponding to a strength

Solution.

Refer to **Fig. 4.4.**

Noting that, $R = \log r$,

$$A_{r,i} = C_i + m_i R = 97.20 - 8.293\,R$$

$$A_{r,f} = C_f + m_f R = 69.20 - 6.282\,R$$

$$A_{r,i} - A_{r,f} = 28.00 - 2.011\,R.$$

Substituting these in the Selfing equation with LHS equated to 29.15

$$A_{r,\,2.167} = \frac{(97.20 - 8.293R) \times (69.20 - 6.282R) \times (3 - 1)}{1.267 \times (28.00 - 2.011R) + 3 \times (69.20 - 6.282R) - 1 \times (97.20 - 8.293R)}$$

$$\Rightarrow \quad 29.15 = \frac{6726.24 - 1184.486R + 52.097R^2}{365.476 - 31.497\,R}$$

$$\Rightarrow \quad 52.097\,R^2 - 266.348R - 3927.385 = 0$$

Solving this quadratic equation in R

$$\Rightarrow \quad R = \frac{266.248 \pm 943.070}{104.194} = 11.607 \text{ (take + ve value only)}$$

$$\Rightarrow \quad \log r = 11.607 \quad \Rightarrow \quad r = 4.049 \text{ (\textbf{actual value} is 4.0)}$$

The **required blend ratio** is **4** at $t = 2.167$ hr.

Example 4.7. *A high strength concrete possesses the following, in the partially setting zone:*

$$t_i = 0.75 \text{ hr},\ t_f = 3.75 \text{ hr}$$

CSP-s (in kg/sq cm) are : $C_i = 583$, $m_i = 15.62$, $C_f = 558$, $m_f = -31.25$, *obtained from the ArkR-plot, where* $R = \log_{10}(r)$.

***Determine** the strength for the composite mix* $A_{r,\,T,}$ *with a blend ratio* $r = 3$, *done at a time lag of 1.5 hr, using time lag parameter* $T = t^n$, *for the case* $n = 1$, *and also for* $n = 0.5$, *and* $n = 2$. ***Compare** the three results.*

Solution. In the time lag parameter T (expressed as t^n), if $n = 1$, for which $T = t$, the strength variations appear as the usual curved shape, (earlier identified as Type I, Type II etc). For $n = 0.5$ and $n = 2$, for which $T = \sqrt{t}$ and $T = t^2$, respectively, the said variations are taken as linear ones (similar to the variations as shown in **Fig. 3.8**).

This problem requires to enquire the effectiveness of the use of the time lag, $T = t$, in comparison to that of the time lag parameter $T = \sqrt{t}$, and $T = t^2$.

Here, $\quad t_i = 0.75$ hr, $t_f = 3.75$ hr , $t = 1.5$ hr

$$\Rightarrow \quad \sqrt{t_i} = 0.8660,\ \sqrt{t_f} = 1.9365,\ \sqrt{t} = 1.2247$$

$$\Rightarrow \quad t_i^2 = 0.5625,\ t_i^2 = 14.0625,\ t^2 = 2.25$$

We recall the second of the Selfing equation (4.3) here as

$$A(r, T) = \frac{A(r, T_i) \times A(r, T_f) \times (T_f - T_i)}{A(r, T_i) \times (T - T_i) + A(r, T_f) \times (T_f - T)}.$$

To find $A_{3,\,1.5}$, (for $r = 3$, and for $n = 1$, $T = t = 1.5$):

Substitute the corresponding data concerned with time lag parameter t, along with

$$A_{ri} = C_i + m_i .\log_{10}(r) = A_{3,\,0.75} = 583 + 15.62 \times \log_{10}(3) = 590.5$$

$$A_{rf} = C_f + m_f .\log_{10}(r) = A_{3,\,3.75} = 558 - 31.25 \times \log_{10}(3) = 543.0908$$

$$A_{3,\,1.5} = \frac{590.5 \times 543.0908 \times (3.75 - 0.75)}{590.5 \times (1.5 - 0.75) + 543.0908 \times (3.75 - 1.5)}$$

$$= \frac{962090.67}{442.875 + 1221.954} = \frac{962090.67}{1664.829} = \mathbf{577.89} \text{ kg/sq cm.}$$

To find $A_{3, 1.2247}$, (for $r = 3$, and for $n = 0.5$, $T = t^{0.5} = 1.2247$):

Substitute the corresponding data concerned with time lag parameter $\sqrt{t}$, along with

$$A_{ri} = C_i + m_i .\log_{10}(r) = A_{3, 0.75} = 583 + 15.62 \times \log_{10}(3) = 590.5$$

$$A_{rf} = C_f + m_f .\log_{10}(r) = A_{3, 3.75} = 558 - 31.25 \times \log_{10}(3) = 543.0908$$

[no change]

$$A_{3, 1.2247} = \frac{590.5 \times 543.0908 \times (1.9365 - 0.867)}{590.5 \times (1.2247 - 0867) + 543.0908 \times (1.9365 - 1.2247)}$$

$$= \frac{342985.32}{211.2219 + 386.7942} = \frac{342985.27}{597.7961} = \mathbf{573.75} \text{ kg/sq cm.}$$

To find $A_{3, 2.25}$, (for $r = 3$, and for $n = 2$, $T = t^2 = 2.25$):

Substitute the corresponding data concerned with time lag parameter t^2, along with

$$A_{ri} = C_i + m_i .\log_{10}(r) = A_{3, 0.75} = 583 + 15.62 \times \log_{10}(3) = 590.5$$

$$A_{rf} = C_f + m_f .\log_{10}(r) = A_{3, 3.75} = 558 - 31.25 \times \log_{10}(3) = 543.0908$$

[no change]

$$A_{3, 2.25} = \frac{590.5 \times 543.0908 \times (14.0625 - 0.5625)}{590.5 \times (2.25 - 0.5625) + 543.0908 \times (14.0625 - 2.25)}$$

$$= \frac{4329408}{996.4688 + 6415.2955} = \frac{4329408}{7411.7643} = \mathbf{584.13} \text{ kg/sq cm.}$$

Thus the **required Selfed strength $A_{3, T}$** is

577.89 kg/sq cm, (for $n = 1$, $T = t = 1.5$)

573.75 kg/sq cm, (for $n = 0.5$, $T = t^{0.5} = 1.2247$)

584.13 kg/sq cm, (for $n = 2$, $T = t^2 = 2.25$).

Comparing the strengths for $n = 0.5$ and $n = 2$, with respect to the strength for $n = 1$, it can be seen that, the strength corresponding to $n = 0.5$ and that for $n = 2$, differ only by – 0.72% and + 1.08%, respectively, which for all practical purpose can be taken as nil.

[**Note:** The negligible percentage of error in the analytical strength $A_{r, T}$, (where, $T = t^n$), for different values on n, indicates clearly, that it will be sufficient enough to go ahead for strength calculation with $T = t$, that is, for $n = 1$. This aspect has been discussed earlier in **Art. 3.4**, item 2.]

Example 4.8. *In a concreting programme with 1:2:4 concrete, the strengths (in kg/sq cm) of two separate blended mixes were interpreted as 191.74 and 180.75, with r and t (in hr) combinations as, (2, 3) and (4, 2), respectively, with the help of the corresponding CSP-s (in kg/sq cm) viz.,* $C_i = 180.6$, $m_i = 29.6$, $C_f = 133.6$, $m_f = 44.4$, *obtained from the ArkR-plot, where* $R = 1/r$.

The two setting times (in hr)are, $t_i = 0.75$ *and* $t_f = 4.00$.

***Determine** the values of the new CSP-s, if the ArkR-plot was used with* $R = \log(1/r)$.

Solution. The ArkR – plot is the linearised representation of the strengths at t_i as well as t_f, with respect to the function R.

This problem requires to obtain the CSP-s from the ArkR-plot using some R- function of choice, when the CSP-s obtained from another ArkR-plot using some other R- function is given.

The zero value of R for the variation $R = 1/r$, is met with for $r = \infty$, where as, for the case of $R = \log(1/r)$, it occurs for $r = 1$. Whatever be the nature of the function R, the intercepts $C_k - s$ (C_i or C_f) actually, are the strength ordinates at $R = 0$. The corresponding slopes $m_k - s$ (m_i or m_f) can be worked out accordingly. (**Fig. 3.8A**).

The **given** strengths (in kg/sq cm) are $A_{2,3} = 191.74$, and $A_{4,2} = 180.75$.

Let the **required** CSP-s for the case $R = \log(1/r)$ are (D_i, n_i) and (D_f, n_f), and the corresponding strengths in this system are $B_{r,t}$.

For the combinations ($r = 2$ and $t = 3$), and ($r = 4$ and $t = 2$), at t_i and t_f :

For $B_{r,k}$ – strengths (with unknowns $D_k - s$ and $n_k - s$)

$$B_{2,i} = D_i + n_i \,.\, \log\left(\frac{1}{2}\right) = D_i - 0.301 \times n_i$$

$$B_{2,f} = D_f + n_f \,.\, \log\left(\frac{1}{2}\right) = D_f - 0.301 \times n_f$$

$$B_{4,i} = D_i + n_i \,.\, \log\left(\frac{1}{4}\right) = D_i - 0.602 \times n_i$$

$$B_{2,i} = D_f + n_f \,.\, \log\left(\frac{1}{4}\right) = D_f - 0.602 \times n_f$$

For $A_{r,k}$ – strengths (with given values of C_k – s and $m_k - s$).

Substitute the given values of C_k and m_k for the $A_{r,k}$ strengths, we have

$$A_{2,i} = C_i + m_i \,.\, \frac{1}{2} = 180.6 + 29.6 \times \frac{1}{2} = 195.4$$

$$A_{2,f} = C_f + m_f \cdot \frac{1}{2} = 133.6 + 44.4 \times \frac{1}{2} = 155.8$$

$$A_{4,i} = C_i + m_i \cdot \frac{1}{4} = 180.6 + 29.6 \times \frac{1}{4} = 188.0$$

$$A_{4,f} = C_f + m_f \cdot \frac{1}{4} = 133.6 + 44.4 \times \frac{1}{4} = 144.7$$

Noting that, the **value of the strength is unique and does not depend on the variation of R-function**, we have four conditions, (two for $r = 2$, and two for $r = 4$, at t_i and t_f), namely

$$B_{2,i} = A_{2,i}$$
$$B_{2,f} = A_{2,f}$$
$$B_{4,i} = A_{4,i}$$
$$B_{4,f} = A_{4,f}$$

Thus
$$B_{2,i} = A_{2,i} \Rightarrow D_i - 0.301 \times n_i = 195.4$$
$$B_{2,f} = A_{2,f} \Rightarrow D_f - 0.301 \times n_f = 155.8$$
$$B_{4,i} = A_{4,i} \Rightarrow D_i - 0.602 \times n_i = 188.0$$
$$B_{4,f} = A_{4,f} \Rightarrow D_f - 0.602 \times n_f = 144.7$$

Solving these four equations

$$D_i = 202.800, n_i = 24.585$$
$$D_f = 166.900, n_f = 36.877$$

These are the **required values of the CSP-s** corresponding the ArkR- plot for which $R = \log(1/r)$.

Example 4.9. *Predict the Selfed strength $A_{6.5,\,2.5}$ (with blend ratio r = 6.5, time lag t = 2.5 hr), using the CSP-s corresponding to R = 1/r, as well as R = log (1/r), and compare the analytical results.*

All other relevant data are same as in Example 4.8.

Solution. This problem refers to checking of CSP-s obtained from one R- function with the corresponding ones from another R- function.

The setting times (in hr) are, $t_i = 0.75$ and $t_f = 4.00$, the time lag at Selfing $t = 2.5$

$$\Rightarrow \quad t_f - t_i = 3.25, t - t_i = 1.75, t_f - t = 1.5$$

In $A_{r,k}$-strength corresponding to $R = 1/r$,

the CSP-s (in kg/sq cm) are, $C_i = 180.6$, $m_i = 29.6$, $C_f = 133.6$, $m_f = 44.4$

Thus $$A_{6.5,\,i} = C_i + m_i \,.\, \frac{1}{r} = 180.6 + 29.6 \times \frac{1}{6.5} = 185.154$$

$$A_{6.5,f} = C_f + m_f \,.\, \frac{1}{r} = 133.6 + 44.4 \times \frac{1}{6.5} = 140.431$$

Using these in (4.4)

$$\frac{t_f - t_i}{A_{r,t}} = \frac{t - t_i}{A_{r,f}} + \frac{t_f - t}{A_{r,i}}$$

$$\Rightarrow \quad \frac{3.25}{A_{6.5,2.5}} = \frac{1.75}{140.431} + \frac{1.5}{185.154}$$

$$\Rightarrow \quad A_{6.5,\,2.5} = \mathbf{150.05} \text{ kg /sq cm.}$$

Thus $$B_{6.5,\,i} = D_i + n_i \,.\, \log \frac{1}{r} = 202.800 + 24.585 \times \log \frac{1}{6.5} = 182.815$$

$$B_{6.5,f} = D_f + n_f . \log \frac{1}{r} = 166.900 + 36.877 \times \log \frac{1}{6.5} = 136.920$$

Using these in (4.4)

$$\frac{t_f - t_i}{B_{r,t}} = \frac{t - t_i}{B_{r,f}} + \frac{t_f - t}{B_{r,i}}$$

$$\Rightarrow \quad \frac{3.25}{B_{6.5,2.5}} = \frac{1.75}{136.923} + \frac{1.5}{182.815}$$

$$\Rightarrow \quad B_{6.5,\,2.5} = \mathbf{154.87} \text{ kg /sq cm.}$$

The two results **differ** by **3.21%.**

Example 4.9A. ***Establish*** *the Generalised Sefing formula (3.33 – 3.35) with the help of data as given in Example 4.9 with the following data:*

Given: *The setting times (in hr) are,* $t_i = 0.75$ *and* $t_f = 4.00$*, the time lag at Selfing* $t = 2.5$, $r = 6.5$.

$$\Rightarrow \quad t_f - t_i = 3.25,\ t - t_i = 1.75,\ t_f - t = 1.5.$$

Solution. This problem requires to get the selfed strength $A_{6.5,2.5}$ with the help of $A_{r,\,t}$ – equation (3.4 or 3.5) using boundary time lags $t_i = 0.75$ and $t_f = 4.00$ first, and then to verify the same strength $A_{6.5,2.5}$ with the help of any two boundary time lags t_p and t_q using (3.33 or 3.34).

Let $t_p = 2$ and $t_q = 3.2$, $t = 2.5$

$\Rightarrow$ $t_q - t_p = 1.2$, $t - t_p = 0.5$, $t_q - t = 0.7$.

We have from **Example 4.9**, (all strengths in kg/sq cm)

$$A_{6.5,\,i} = A_{6.5,\,0.75} = 185.154,\ A_{6.5,\,f} = A_{6.5,\,4} = 140.431$$

To find A (6.5, 2.5) for $r = 6.5$, and $t = 2.5$ hr, and $t_i = 0.75$ hr, $t_f = 4$ hr, using (3.4) or (4.3)

$$A_{6.5,2.5} = \frac{185.154 \times 140.431 \times (4 - 0.75)}{185.154 \times (2.5 - 0.75) + 140.431 \times (4 - 2.5)}$$

$$= \frac{84504.42445}{534.666} = \mathbf{158.051.}$$

To find A (6.5, 2) and A (6.5, 3.2) for $r = 6.5$, and $t_p = 2$ hr , and $t_q = 3.2$ hr, using (3.5) or (4.4)

$$\frac{3.25}{A_{6.25,2}} = \frac{2 - 0.75}{A_{6.5,4}} + \frac{4 - 2}{A_{6.5,0.75}} \quad \Rightarrow \quad \frac{3.25}{A_{6.5,2}} = \frac{2 - 0.75}{140.431} + \frac{4 - 2}{185.154}$$

$$\Rightarrow \quad A_{6.5,2} = 164.975$$

$$\frac{3.25}{A_{6.25,3.2}} = \frac{3.2 - 0.75}{A_{6.5,4}} + \frac{4 - 3.2}{A_{6.5,0.75}} \quad \Rightarrow \quad \frac{3.25}{A_{6.25,3.2}} = \frac{3.2 - 0.75}{140.431} + \frac{4 - 3.2}{185.154}$$

$$\Rightarrow \quad A_{6.5,3.2} = 149.288.$$

To verify whether using (3.34), A(6.5, 2.5) = 150.05, or not:

Substitute $t_p = 2$ and $t_q = 3.2$, $t = 2.5$, so that, $t_q - t_p = 1.2$, $t - t_p = 0.5$, $t_q - t = 0.7$, in (3.5) to give:

$$\frac{3.2 - 2}{A_{6.5,2.5}} = \frac{2.5 - 2}{A_{6.5,3.2}} + \frac{3.2 - 2.5}{A_{6.5,2}} \quad \Rightarrow \quad \frac{1.2}{A_{6.25,2.5}} = \frac{3.2 - 0.75}{149.288} + \frac{4 - 3.2}{164.975}$$

$$\Rightarrow \quad A_{6.5,\,2.5} = \mathbf{158.103.}$$

Alternatively, using (3.33)

$$A_{6.5,\,2.5} = \frac{164.975 \times 149.288 \times (3.2 - 2)}{164.975 \times (2.5 - 2) + 149.288 \times (3.2 - 2.5)}$$

$$= \frac{29554.5454}{186.9891} = \mathbf{158.055.}$$

Alternatively, to get this strength using (3.11)

$$A_{r,\,i} = 185.154,\ A_{r,f} = 140.431 \quad \Rightarrow \quad a_{r,\,i} = 0.005401,\ a_{r,f} = 0.007121$$

$t_f = 4$, $t_i = 0.75$, $t = 2.5$, $A_{r,t}$ (at $t = 2.5$) = **to be determined.**

$$\frac{0.007121 - a_{rt}}{a_{rt} - 0.005401} = \frac{4 - 2.5}{2.5 - 0.75} \Rightarrow a_{r,t} = 0.006327 \Rightarrow A_{r,t} = \mathbf{158.05.}$$

The **resultant strength** ***A*** **(6.5, 2.5) calculated using Selfing formula and its Generalised version is same** ***viz.*****, 158.01 kg/sq cm.**

Example 4.10. *A composite fibre reinforced concrete of M40 grade was prepared with different percentages (p) of steel fibres (of crimped shape) by weight of concrete (viz., 0.0, 1.5, 3.0 and 6.0), and of different aspect ratios (= l/d = stretched length to diameter ratio as 38, and 51). Selfing experiments were conducted on these mixes by blending freshly prepared mix with the basic partially set ones.*

The pull-out bond test on a particular mix type A, is expressed as function of the fibre percentage p (with $\lambda = 38$), and time lag t, viz., A(p, t), was carried out at different time lags t (in hr). The pull-out bond strength vs time lag variations were of Type II variety, and the strength $A_{p,k}$, (k is either 0, i or f, indicating $t = 0$, t_i or t_f, respectively), as the function of fibre percentage p, was derived as

$$A_{p,k} = (1 + p)^{m_k} \cdot C_k \qquad (4.7)$$

where, the CSP-s for the Zone I are: (C_k – s and m_k – s), viz., ($C_0 = 6.92$, $m_0 = 0.145$), ($C_i = 7.29$, $m_i = 0.133$), and for t =0 and t_i, respectively.

Using analogously the selfing equation (3.19) on replacing the blend ratio r by the fibre percentage p, ***predict*** *the bond strength $A_{3,2}$ (in MPa) for the blended partially set fibre reinforced concrete for the combination (p = 3, t = 2), if $t_i = 3$ hr, and $t_f = 4$ hr.*

Compare *the predicted strength with the experimental one.*

Solution. This problem is the illustration of the analogous application of the Selfing equation, and belongs to the Category 2, as described in **Art 1.5.**

Since the strength *vs* time lag variations conforms to Type II category, we recall (3.19) here as

for **Zone I**:
$$A_{p,t} = \frac{A_{p,0} \cdot A_{p,i} \cdot (t_i - 0)}{t \cdot A_{p,0} + A_{p,i} \cdot (t_i - t)}$$

for **Zone II**:
$$A_{p,t} = \frac{A_{p,i} \cdot A_{p,f} \cdot (t_f - t_i)}{t \cdot (A_{p,i} - A_{p,f}) + A_{p,f} \cdot t_f - A_{p,i} \cdot t_i} \qquad (4.8)$$

[Solution-wise, the rest of the procedure is same as for Example 4.4.]

To find ***A*****(3, 2) for** ***p*** **= 2, and at** ***t*** **= 1.5 hr:**

The time lag of observation at $t = 2$ hr, falls within Zone *I*, *i.e.*, between $t = 0$ to $t = t_i = 3$ hr.

$$A_{p,0} = (1+p)^{m_0}.C_0 \quad \Rightarrow \quad A_{3,0} = (1+3)^{0.145} \times 6.92 = 8.46$$

$$A_{P,i} = (1+P)^{m_i}.C_i \quad \Rightarrow \quad A_{3,3} = (1+3)^{0.133} \times 7.29 = 8.77$$

$$A_{3,2} = \frac{8.46 \times 8.77 \times 3}{8.46 \times 2 + 8.77 \times (3-2)} = 8.66 \text{ MPa}$$

The **experimental** value was = **8.52** MPa , the **difference** = **1.62** %.

Example 4.11. *Impact strength (in Nm) was measured just at the appearance of first crack* f_{iml}, *for the composite partially set fibre reinforced concrete of M20 grade, with steel fibre (of* $\lambda = 38$*), 6 percent by weight of concrete. If the CSP–s (viz.,* C_k *– s and* m_k *–s), for the ZONE II are: (*$C_i = 460$, $m_i = 0.664$*), and (*$C_f = 241$, $m_f = 0.875$*), at* $t_i = 3$ *hr and* $t_f = 4$ *hr, respectively,* ***calculate*** *the impact strength of the blended fibre reinforced concrete done at the time lag of 3.5 hr, and* ***compare*** *with that of the experimental one.*

Solution. This problem is the illustration of the analogous application of the Selfing equation, and belongs to the Category 2, as described in **Art 1.5.**

[Solution-wise, the rest of the procedure is same as for Example 4.10.]

Since the strength *vs* time lag variations conforms to Type II category, we recall (4.8) here for Zone II .

To find *A*(6, 3.5) for *p* = 6, and at *t* = 3.5 hr:

The time lag of observation at $t = 3.5$ hr, falls within Zone II, *i.e.*, between $t = t_i = 3$ hr, to $t = t_f = 4$ hr.

$$A_{p,i} = (1+p)^{m_i}.C_i \quad \Rightarrow \quad A_{6,3} = (1+6)^{0.664} \times 460 = 1674.57$$

$$A_{p,f} = (1+p)^{m_f}.C_f \quad \Rightarrow \quad A_{6,4} = (1+6)^{0.875} \times 241 = 1322.75$$

$$A_{6,3.5} = \frac{1674.57 \times 1322.75 \times (4-1)}{1674.57 \times (3.5-3) + 1322.75 \times (4-3.5)} = \mathbf{1478} \text{ Nm}$$

The **experimental** value was = **1468** Nm, the **difference** = **0.68** %.

4.3 SELFING INVOLVING TWO MIXES (GENERAL SELFING)

Example 4.12. *A blended composite concrete was prepared by blending two individual blended partially set mixes* A_O *and* A_F *of the same mix type A, in proportion 2:1.*

The relatively old partially set Pure Selfed mix A_O*, again was prepared separately earlier by blending basic THC and fresh IPC in equal quantities at a time lag of 1 hr.*

The relatively fresh partially set pure selfed mix A_F, was also prepared separately earlier by adding older THC and fresher quantities IPC in 4:1 ratio, cast at a delayed time of 1 hr.

If the two setting times (in hr) of the mix A are 0.75 and 4.00, and the CSP-s (in kg /sq cm) in the A_{rk} vs 1/r - plot, are C_i = 180.6, m_i = 29.6, C_f = 133.6, m_f = 44.4, ***predict*** *the final strength of the resultant composite mix.*

Solution. This problem requires to find the values of the blended strengths relating to (r, t) combinations as: (r_1, t_1) and (r_2, t_2), when the CSP-*s* are given.

The mix A_O is the older partially set mix. Its $r_1 = 1$, and $t_1 = 2$,

So that its strength is represented as $A(r_1, t_1) = A(1, 2) = A_{1, 2.}$

It was prepared, in turn, through pure selfing by blending basic THC with the IPC of the mix type *A*.

This is represented by **split** as: $A_{1, 2} = [(A_{\infty, 2}) / (A_{0, 2})]_1$

The mix A_F is the fresher partially set mix. Its $r_2 = 4$, and $t_2 = 1$,

So that its strength is represented as $A(r_2, t_2) = A\ (4, 1) = A_{4,1}$.

It was prepared, in turn, through pure selfing by blending basic THC with the IPC of the mix type *A*.

This is represented by **split** as: $A_{4, 1} = [(A_{\infty, 1}) / (A_{0, 1})]_4$.

The General Selfing representation of the two partially set mixes **(Fig. 4.5)** is given by

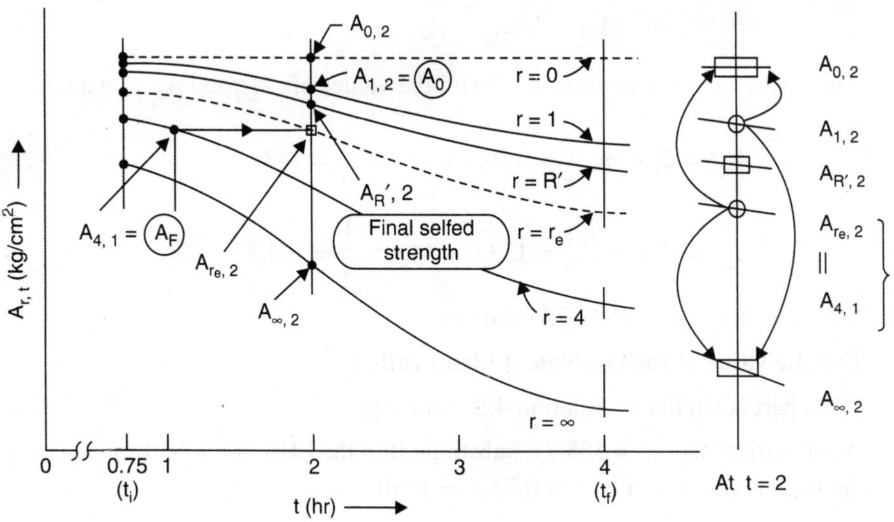

Fig. 4.5. General selfed strength due to blending

$[(A(r_1, t_1)) / (A(r_2, t_2))]_r \Rightarrow [(A_{1,\,2}) / (A_{4,\,1})]_2$

$\Rightarrow \quad [((A_{\infty,\,2}) / (A_{0,\,2}))_1 / ((A_{\infty,\,1}) / (A_{0,\,1}))_4]_2$

To read final strength, select suitable time lag:

Select that the resultant selfed strength of the two mixes A_O and A_F is to be read on the ordinate for the larger value of the time lags (*i.e.*, between $t_2 = 1$ hr and $t_1 = 2$ hr), *viz.*, at $t_1 = 2$ hr. Then it requires **to find** the required resultant blend ratio R', for which the Blend ratio equation (3.30), may be made use of. Then the final strength is read as $A(R', 2)$ or $A_{R',\,2}$.

To get blend ratio r_e equivalent to r_2, on t_1:

For the value of the fresher strength $A(4, 1)$:

Project the fresher strength $A(4, 1)$ on the ordinate at $t_1 = 2$, such that the strength $A(4, 1)$ is equal to $A(r_e, 2)$ with the blend ratio r_e, which is yet **to be determined**. Then using Selfing formula (4.4), r_e is expressed in terms of r_2 (= 2), as

$$\frac{t_f - t_i}{A_{r,t}} = \frac{t - t_i}{A_{r,f}} + \frac{t_f - t}{A_{r,i}}$$

Here, $t = 1$, $t_i = 0.75$, $t_f = 4.00$

$$\Rightarrow \quad \frac{4 - 0.75}{A_{4,1}} = \frac{1 - 0.75}{A_{4,f}} + \frac{4 - 1}{A_{4,i}}$$

$$\Rightarrow \quad \frac{3.25}{A_{4,1}} = \frac{0.25}{A_{4,f}} + \frac{3}{A_{4,i}}$$

This expression when substituted with the values of $A_{4,\,i}$ and $A_{4,\,F}$, namely

$$A_{4,\,i} = C_i + m_i \cdot \frac{1}{r} = 180.6 + 29.6 \times \frac{1}{4} = 188$$

$$A_{4,\,f} = C_f + m_f \cdot \frac{1}{r} = 133.6 + 44.4 \times \frac{1}{4} = 144.7$$

yields $A(4, 1) = A(r_{e,\,2}) = 183.77$ kg /sq cm.

For the value of the equivalent blend ratio r_e:

[This part is similar to Solution 4.5, item (*c*)]

After getting $A(r_{e,\,2}) = 183.77$, substitute it to the LHS of the Selfing equation (4.3). The RHS requires, (with , $t_i = 0.75$, $t_f = 4.00$)

$$A(r_e, t_i) = A(r_e, 0.75) = C_i + m_i \cdot \frac{1}{r_e} = 180.6 + 29.6 \times \frac{1}{r_e} = 180.6 + 29.6y$$

$$A(r_e, t_f) = A(r_e, 4.00) = C_f + m_f \cdot \frac{1}{r_e} = 133.6 + 44.4 \times \frac{1}{r_e} = 133.6 + 44.4y$$

where, $y = 1/r_e$.

$$\text{Thus } 183.77 = \frac{(180.6 + 29.6y) \times (133.6 + 44.4y) \times (4 - 0.75)}{2 \times (47 - 14.8y) + 4 \times (133.6 + 44.4y) - 0.75 \times (180.6 + 29.6y)}$$

$$\Rightarrow \quad 183.77 = \frac{4271.28y^2 + 38912.9y + 78416.52}{125.8y + 492.95}$$

$$\Rightarrow \quad 4271.28y^2 + 15794.63y - 12172.90 = 0$$

$$\Rightarrow \quad \frac{1}{r_e} = y = \frac{-15794.63 \pm \sqrt{249470336.8 + 207975457.2}}{8542.56}$$

$$= \frac{-15794.63 \pm 21387.98}{8542.56}$$

Omitting the – ve value

$$\Rightarrow \quad \frac{1}{r_e} = 0.65476 \quad \Rightarrow \quad r_e = 1.527$$

The **equivalent blend ratio** is $r_e = 1.527$.

For resultant blend ratio R':

For getting the final resultant General Selfed strength $A(R', 2)$, first calculate the value for R' using (3.30).

With $r_1 = 1$, $r_e = 1.527$ and $r = 2$,

$$R' = \frac{r_e(1 + r_1 + rr_1) + rr_1}{1 + r + r_1 + rr_e} = \frac{1.527 \times (1 + 1 + 2) + 2}{1 + 2 + 1 + 1.527 \times 2} = 1.149, \text{ (say } = 1.15)$$

For Resultant General Selfed strength $A(R', 2)$:

$$A(R', t_1) = [(A(r_1, t_1)) / (A(r_2, t_2))]_r \quad \Rightarrow \quad A_{1.15,\,2} = [(A_{1,\,2}) / (A_{4,\,1})]_2$$

$$\Rightarrow \quad A_{1.15,\,2} = [(A(1, 2))/(A(r_e, 2))]_2 \quad \Rightarrow \quad A_{1.15,\,2} = [(A_{1,\,2}) / (A_{1.527,\,2})]_2$$

Following the same procedure as followed for calculating $A_{4,\,1}$ in this Example, we have

$$A_{1.15,\,i} = C_i + m_i \cdot \frac{1}{r} = 180.6 + 29.6 \times \frac{1}{1.15} = 206.34$$

$$A_{1.15,\,f} = C_f + m_f \cdot \frac{1}{r} = 133.6 + 44.4 \times \frac{1}{1.15} = 172.21$$

Using these in (4.4) with $t = 2$, $t_i = 0.75$, $t_f = 4.00$

$$\frac{t_f - t_i}{A_{r,t}} = \frac{t - t_i}{A_{r,f}} + \frac{t_f - t}{A_{r,i}}$$

$$\Rightarrow \qquad \frac{3.25}{A_{1.15,2}} = \frac{2 - 0.75}{172.21} + \frac{4 - 2}{206.34}$$

$$\Rightarrow \qquad \mathbf{A_{1.15,\,2} = 191.73.}$$

The **required final general selfed strength** is predicted as **191.73** kg /sq cm.

Example 4.13. *The strengths (in kg /sq cm) of two individual partially set selfed mixes of the same mix type A, are196.53 and 183.77, observed for the (r, t) combinations of (1, 2) and (4, 1), respectively. The strengths at the two setting times, $t_i = 0.75$ hr and $t_f = 4.00$ hr of the basic THC, are 180.6 and 133.6, respectively.* ***Determine*** *the two slopes, m_i and m_f of the linear plot of the strength vs 1/r.*

Solution. This problem requires to find the two unknown CSP-*s*, (out of four CSP-*s*), when two strengths are given. This Example may be considered as the reverse problem of the Example 4.10.

In the ArkR - plot, the strength at $R = 0$ gives the intercept C_k. It occurs for the strength curve, $r = \infty$, which actually corresponds to the basic mix. Thus

$$A(\infty, t_i) = C_i = 180.6, \quad \text{and} \quad A(\infty, t_f) = C_f = 133.6$$

The **knowns** are, C_i and C_f, and the **unknowns** to be determined are, m_i and m_f.

Refer to **Fig. 4.6.**

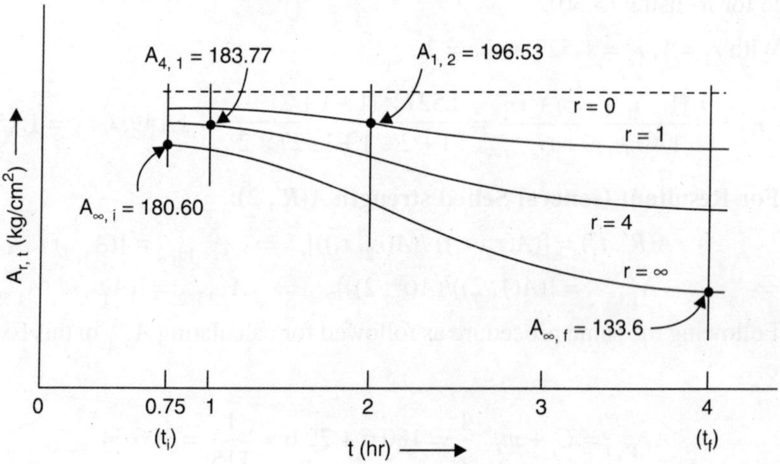

Fig. 4.6. CSP-s corresponding to R

To calculate $A_{1,2}$:

$$A_{1,i} = C_i + m_f \cdot \frac{1}{r} = 180.6 + m_i \times \frac{1}{1} = C_i + m_i$$

$$A_{1,f} = C_f + m_f \cdot \frac{1}{r} = 133.6 + m_f \times \frac{1}{1} = C_f + m_f$$

With $t = 2$, $r = 1$ and $A_{1,2} = 196.53$, we have

$$\frac{t_f - t_i}{A_{r,t}} = \frac{t - t_i}{A_{r,f}} + \frac{t_f - t}{A_{r,i}}$$

$$\frac{3.25}{A_{1,2}} = \frac{2 - 0.75}{C_f + m_f} + \frac{4 - 2}{C_i + m_i}$$

$$\Rightarrow \qquad \frac{3.25}{196.52} = \frac{1.25}{133.6 + m_f} + \frac{2}{180.6 + m_i}$$

$$\Rightarrow \qquad \frac{1.25}{133.6 + m_f} + \frac{2}{180.6 + m_i} = 0.01654$$

Dividing both sides by 0.01654

$$\Rightarrow \qquad \frac{75.5043}{133.6 + m_f} + \frac{120.8068}{180.6 + m_i} = 1$$

$$\Rightarrow \qquad \frac{75.5043}{133.6 + m_f} = 1 - \frac{120.8068}{180.6 + m_i} = \frac{59.7932 + m_i}{180.6 + m_i}$$

Taking reciprocal of both sides

$$\Rightarrow \qquad \frac{133.6 + m_f}{75.5043} = \frac{180.6 + m_i}{59.7932 + m_i}$$

$$\Rightarrow \qquad 1.7694 + 0.01324 m_f = 1 + \frac{120.8068}{m_i + 59.7932}$$

$$\Rightarrow \qquad 58.1118 + m_f = \frac{9124.3807}{m_i + 59.7932} \qquad (4.9a)$$

To calculate $A_{4,1}$:

$$A_{4,i} = C_i + m_i \cdot \frac{1}{r} = 180.6 + m_i \times \frac{1}{4} = 180.6 + 0.25m_i$$

$$A_{4,f} = C_f + m_f \cdot \frac{1}{r} = 133.6 + m_f \times \frac{1}{4} = 133.6 + 0.25m_f$$

With $t = 1$, $r = 4$ and $A_{4,1} = 183.77$, we have

$$\frac{t_f - t_i}{A_{r,t}} = \frac{t - t_i}{A_{r,f}} + \frac{t_f - t}{A_{r,i}}$$

$$\frac{3.25}{A_{4,1}} = \frac{1 - 0.75}{C_f + m_f + \frac{1}{4}} + \frac{4 - 1}{C_i + m_i \times \frac{1}{4}}$$

$$\Rightarrow \quad \frac{3.25}{183.77} = \frac{0.25}{133.6 + 0.25m_f} + \frac{3}{180.6 + 0.25m_i}$$

$$\Rightarrow \quad \frac{0.25}{133.6 + 0.25m_f} + \frac{3}{180.6 + 0.25m_i} = 0.01769$$

Dividing both sides by 0.1769

$$\Rightarrow \quad \frac{14.1323}{133.6 + 0.25m_f} + \frac{169.5873}{180.6 + 0.25m_i} = 1$$

$$\Rightarrow \quad \frac{14.1323}{133.6 + 0.25m_f} = 1 - \frac{169.5873}{180.6 + 0.25m_i} = \frac{11.0127 + 0.25m_i}{180.6 + 0.25m_i}$$

Taking reciprocal of both sides

$$\Rightarrow \quad \frac{133.6 + 0.25m_f}{14.1323} = \frac{180.6 + 0.25m_i}{11.0127 + 0.25m_i}$$

$$\Rightarrow \quad 9.4535 + 0.01769m_f = 1 + \frac{169.5873}{0.25m_i + 11.0127}$$

$$\Rightarrow \quad 477.8689 + m_f = \frac{9586.6196}{0.25m_i + 11.0127} \qquad (4.9b)$$

Subtract (4.9*a*) from (4.9*b*) to eliminate m_f

$$419.7571 = \frac{7305.5244m_i + 4727559.3556}{0.25m_i^2 + 25.9681m_i + 658.5176}$$

$$\Rightarrow \qquad 1 = \frac{17.4042m_i + 126.2680}{0.25m_i^2 + 25.9681m_i + 658.5176}$$

$$\Rightarrow \qquad 0.25m_i^2 + 8.5575m_i - 467.7504 = 0$$

$$\Rightarrow \qquad m_i^2 + 34.2300\,m_i - 1871.0016 = 0$$

The roots of this quadratic equation are

$$m_i = \frac{-34.23 \pm \sqrt{34.23^2 + 4 \times 1871.0016}}{2} = \frac{-34.23 \pm 93.0306}{2}$$

$m_i = 29.403$ (accepting +ve value only)

Inserting this value of m_i in (4.9a)

$$m_f = \frac{9124.3807}{29.403 + 59.7962} - 58.1118 = 44.18$$

The **required slopes** are determined as (values within parenthesis are the given data of Example 4.10):

m_i = 29.403 (29.6), m_f = 44.180 (44.4).

4.4 CROSSING INVOLVING TWO MIXES (GENERAL CROSSING)

Example 4.14. *A partially set basic mix of M40 grade concrete at t = 3 hr is required to be blended with some freshly prepared M55 grade concrete at that instant of time lag in equal proportion by weight. The curing condition of the blended mix is of Improper Type viz., $W_{14}A_{14}$, (i.e., out of 28 days of full water curing, first 14 days in water, and next balance of 14 days in air). The time lag parameter $T = \sqrt{t}$, and in ArkR-plot, R = log10 (r).*

*The other data is as **given**:*

Old mix is A (M40), t_i = 0.75 hr, t_f = 3.75 hr

$A(\infty, t_i)$ = 470 kg/sq cm, $A(\infty, t_f)$ = 332 kg/sq cm

Fresh mix is B (M55), t_i = 0.75 hr, t_f = 3.75 hr

B(0, 0) = 568 kg/sq cm, $B(\infty, t_i)$ = 619 kg/sq cm, $B(\infty, t_f)$ = 542 kg/sq cm

CSP-s of mix B : C_i = 604, m_i = 14.58, C_f = 578, m_f = – 33.33

***Predict** the resultant strength of the composite mix.*

Solution. This problem illustrates the prediction of resultant strength in Crossing, since the two mixes in blending are of different types, one is partially set M40 concrete (THC) of strength $A(\infty, 3)$, and the other is fresh M55 concrete (IPC) of strength $B(0, 0) = B(0, 3)\ B(0,T)$

Refer to **Fig. 4.7.**

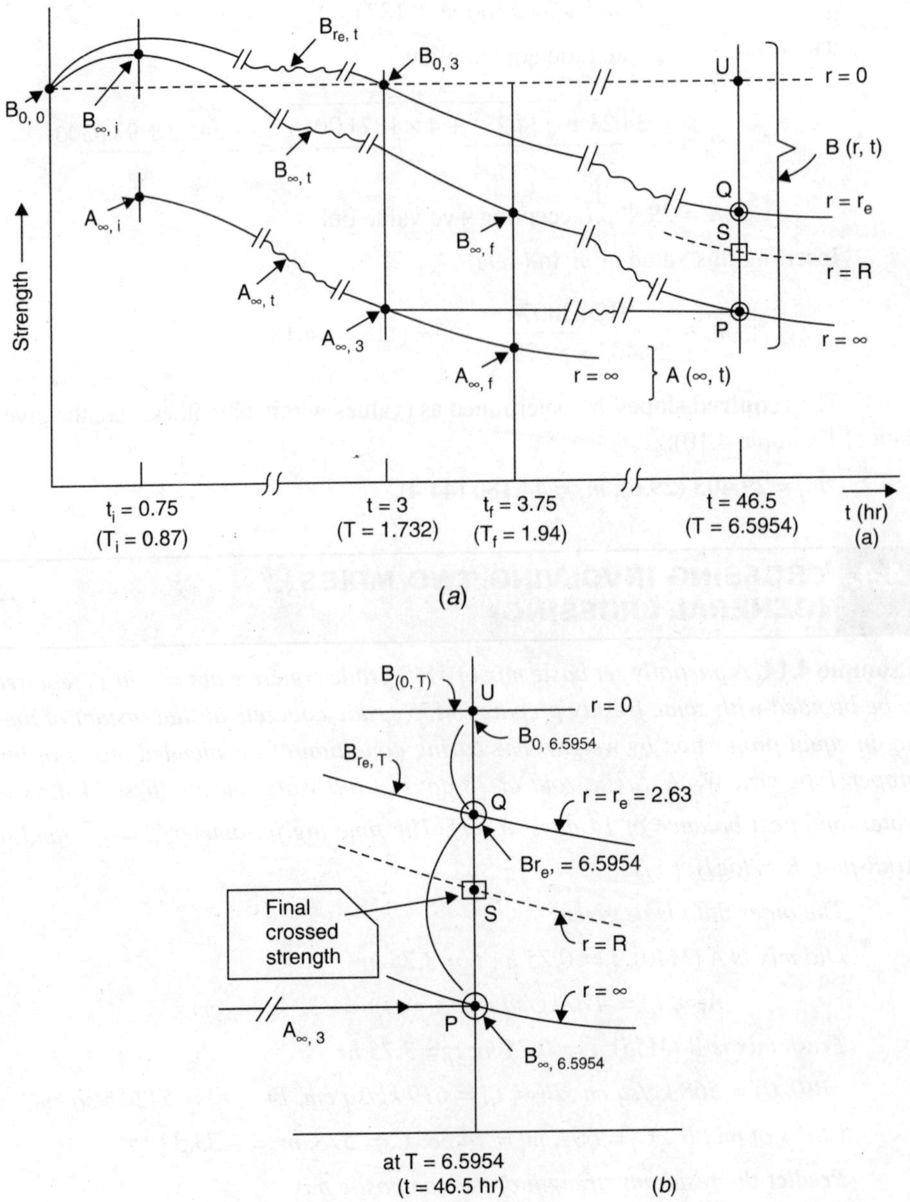

Fig. 4.7. Crossed strength of two mixes

In symbolic representation the entire mathematical operation can be written as

$$C_{R,T} = [(A(\infty, 3)/B\ (0, 0)]1$$

where, $C_{R,T}$ represents the resultant crossed strength, which will be equivalent to that of a third mix C, and is intended to be read corresponding to its blend ratio R (**to be determined**), and time lag parameter T (**if decided to choose**)

Decided that the mix C be the mix B itself.

Then, first it is to **find** the equivalent strength of the old mix $A(\infty, 3)$ corresponding to that of the mix B.

Since M55 concrete shows larger strengths than M40 concrete, it is preferable to **transfer** strength of the old mix $A(\infty, 3)$ in terms of equivalent strength of B. For doing this, the horizontal line from $A(\infty, 3)$ first cuts the strength - time lag curves for $r = \infty$ for the mix B, and this will happen at time lag parameter T.

That is, $A(\infty, 3) = B(\infty, T)$.

Now, **once both the constituent mixes in Crossing are expressed in terms of the same mix type B, the problem reduces to that of Selfing,** namely

$$B_{R,T} = [(B(\infty, T)\ /B(0, 0)]_1,\ \text{(all time lags are with } n = 0.5,\ i.e.,\ T = \sqrt{t}).$$

Rest of the solution procedure is exactly identical to that of the Example 4.12

Time lag parameters:

$$T_i = \sqrt{0.75} = 0.866,\ T_f = \sqrt{3.75} = 1.936,\ T_f - T_i = 1.070$$

$$A(\infty, 3) = \frac{A(\infty, T_i)\,.\,A(\infty, T_f) \times (T_f - T_i)}{(\sqrt{3} - T_i) \times A(\infty, T_i) + (T_f - \sqrt{3}) \times A(\infty, T_f)}$$

$$= \frac{470 \times 332 \times 1.07}{0.866 \times 470 + 0.204 \times 332} = 351.56$$

Hence, $B(\infty, T) = 351.56$.

To find T (**our choice** of time lag parameter where Crossed strength is to be read), use the Selfing equation with LHS $= B(\infty, T) = A(\infty, 3) = 351.56$, and with given values of the mix B (M55) at the two setting times.

$$351.56 = \frac{B(\infty, T_i) \times B(\infty, T_f) \times (T_f - T_i)}{(T - T_i) \times B(\infty, T_i) + (T_f - T) \times B(\infty, T_f)}$$

$$= \frac{619 \times 542 \times 1.07}{(T - 0.866) \times 619 + (1.936 - T) \times 542}$$

Solve it to get, $T = 6.5954\ (\text{hr})^{0.5} \Rightarrow$ the corresponding **time lag t = 43.5** hr.

Note: The time lag at which the $A(\infty, 3)$ of the Mix A is made equivalent to $B(\infty, T)$ of the Mix B, is at an unbelievable elapsed time of 43.5 hr, and by that time no mix remains workable to allow blending. However, it is assumed theoretically that, the strength - time lag curves of mixes continue to maintain their variations indefinitely, if extrapolated. Thus

$$B_{R,T} = [B(\infty, T) / B(0, 0)]_1$$

Again, since the old strength $B(\infty, T)$ is transferred on time lag ordinate T, the fresh strength **needs to be transferred** on the same time lag ordinate, so that the final Crossed strength can be read on this T-ordinate as $B_{R,T}$, where, $T = 6.5954$ $[(\text{hr})^{0.5}]$.

Let $\qquad B(0, 0) = B(r_e, T)$

With **given** value of $B(0, 0) = 568$, and $T = 6.5954$ (**just determined**), find r_e using again the same Selfing equation. Thus

$$568 = \frac{B(r_e, T_i) \times B(r_e, T_f) \times (T_f - T_i)}{(6.5954 - T_i) \times B(r_e, T_i) + (T_f - 6.5954) \times B(r_e, T_f)}$$

Substitute in it

$$B(r_e, T_i) = C_i + m_i \log(r_e) = 604 + 14.58 \log(r_e)$$

$$B(r_e, T_f) = C_f + m_f \log(r_e) = 578 - 33.33 \log(r_e),$$

and **solve** for r_e, to get $\mathbf{r_e = 2.63}$

The resultant Crossed strength expression now is

$$B_{R,T} = [B(\infty, T) / B(r_e, T)]_1 = [B(\infty, T) / B(2.63, T)]_1, \text{ where } T = 6.5954.$$

To get required resultant blend ratio R, substitute $r = 1$ and $r_e = 2.63$, in (3.32)

$$R = r(1 + r_e) + r_e = 1 \times (1 + 2.63) = 2.63 = 6.26$$

The Crossed strength, again from the Selfing equation, is obtained as

$$B(R, T) = \frac{B(R, T_i) \times B(R, T_f) \times (T_f - T_i)}{(6.5954 - T_i) \times B(R, T_i) + (T_f - 6.5954) \times B(R, T_f)}$$

where $\quad R = 6.26,$

$$B(R, T_i) = C_i + m_i \log(R) = 604 + 14.58 \log(6.26) = 615.00$$

$$B(R, T_f) = C_f + m_f \log(R) = 578 - 33.33 \log(6.26) = 551.45$$

Thus with these substitutions

$$B_{R,T} = B_{6.26,6.5954} =$$

$$\frac{B(6.26, 0.866) \times B(6.26, 1.936) \times (1.936 - 0.866)}{(6.5954 - 0.866) \times B(6.26, 0.866) + (1.936 - 6.5954) \times B(6.26, 1.936)}$$

$$= \frac{B(6.26, 0.866) \times B(6.26, 1.936) \times (1.936 - 0.866)}{(6.5954 - 0.866) \times B(6.26, 0.866) + (1.936 - 6.5954) \times B(6.26, 1.936)}$$

$$= \frac{615 \times 551.45 \times 1.07}{(5.729) \times (615.61) - (4.659) \times (551.45)} = 371.51$$

The **analytical resultant Crossed strength is then = 371.51** kg/sq cm.

The **predicted** value **differs** from that of the experimental one by **7.9** % .

Example 4.15. *To a basic concrete mix M20 which has undergone partial setting up to a time lag of 2.5 hr, freshly prepared water cement slurry (w/c ratio = 1.5) is blended with a certain percent by weight of concrete. If the strength (in MPa) of the fresh slurry and that of the finally crossed slurry mixed concrete are 1.25 and 12.52 respectively,* ***determine*** *the required percentage of the blend ratio.*

Solution. This problem of Crossing requires to transfer the strength of the fresh slurry $A_{0,0}$ equivalent to that of the M20 concrete $B_{r,\,t}$ on selecting proper r and t. Since the slurry strength is very low, its strength equivalent on $r = \infty$ for mix B is preferable, and then calculate t.

The Crossing problem then reduces to General Selfing, the strength of which is written as

$$B(R, t) = (B(\infty, 2.25) \,/\, B(\infty, t))_r$$

(in which, R is the final blend ratio on the t–ordinate, so that the final Selfed strength of the slurry mixed composite will be read from the set of $B_{r,\,t}$–variations.

On this t–ordinate thus found, $B(\infty, 2.25)$ [the strength of the partially set basic concrete M20 termed as **old**], is to be projected to get $B(r_e, t)$. Thus the General Selfing with $B(\infty, t)$ as the **fresh** part, further reduces to

$$B(R, t) = ((B(r_e, t) \,/\, B(\infty, t))$$

From this, we **use** $R = r_e + r_e\, r + r$, to get the required proportion of the blend ratio r.

Note: the strength-time variation of the cement slurry follows the trend as shown in **Fig. 3.7 (type V)**. As time lag increases, the strength also increases, the so-called initial setting is largely delayed, and soon after that, the final setting occurs. This behaviour has been noted for all varieties of cement paste and cement slurry, and for all combination of blend ratios (considered for experimental programme in the laboratory).

In the present case, at time lags t (hr) = 0.00, 0.75, 1.50, 2.25, 3.00, and 3.75, the strengths (in MPa) of the cement slurry (call **old** mix A) are : 1.25, 1.44, 1.70, 2.07, 2.66, and 1.70, respectively.

Use **time lag parameter T** (instead of time lag t, use $T = \sqrt{t}$),

In **mix *A* (fresh cement-slurry)**, the data **given**:

The setting times are $t_i = 3$, $t_f = 3.75$, then $T_i = 1.732$, $T_f = 1.936$.

The strength (in MPa) is : $A_{0,0} = 1.25$.

In **mix *B* (M20 concrete)**, the data **given**:

The setting times are : $t_i = 0.75$, $t_f = 3.75$, then $T_i = 0.866$, $T_f = 1.936$, $T_f - T_i =$ 1.07

The strengths (in MPa) are : $B(0,0) = 20.5$, $B(\infty, T_i) = 27.5$, $B(\infty, T_f) = 13.5$.

The CSP-s are : $C_i = 24.5$, $m_i = 3.75$, $C_f = 21.1$, $m_i = -4.166$.

To find strength of the time lag parameter *T* when the fresh slurry is made equivalent to mix *B* at $r = \infty$,

Now $A(0,0) = B(\infty, T)$, where T is **to be determined**.

$$1.25 = \frac{B(\infty, T_i) \, . \, B(\infty, T_f) \times (T_f - T_i)}{(T - T_i) \times B(\infty, T_i) + (T_f - T) \times B(\infty, T_f)}$$

$$= \frac{27.5 \times 13.5 \times 1.07}{(T - 0.866) \times 27.5 + (1.936 - T) \times 13.5} = \frac{397.2375}{14T + 2.321}$$

Solving, $T = 6.1207$, (*i.e.*, $t = 37.463$ hr).

To find the equivalent blend ratio r_e, when $B(\infty, \sqrt{2.25}) = B(r_e, T)$

Find first the basic strength $B(\infty, \sqrt{2.25})$,

$$B(\infty, \sqrt{2.25}) = \frac{B(\infty, T_i) \, . \, B(\infty, T_f) \times (T_f - T_i)}{(T - T_i) \times B(\infty, T_i) + (T_f - T) \times B(\infty, T_f)}$$

$$= \frac{27.5 \times 13.5 \times 1.07}{(1.5 - 0.866) \times 27.5 + (1.936 - 1.5) \times 13.5} = 17.034$$

Thus, we have, $B(\infty, \sqrt{2.25})$ *i.e.*, $B(\infty, 1.5) = 17.034$

To get $B(r_e, T)$, we **require** [putting $X = \log_{10}(r_e)$]

$B(r_e, T_i) = C_i + m_i \log(r_e) = 24.5 + 3.75 \log(r_e) = 24.5 + 3.75\, X$

$B(r_e, T_f) = C_f + m_f \log(r_e) = 21.1 - 4.166 \log(r_e) = 21.1 - 4.166\, X$

$$B(r_e, T) = \frac{B(r_e, T_i) \, . \, B(r_e, T_f) \times (T_f - T_i)}{(T - T_i) \times B(r_e, T_i) + (T_f - T) \times B(r_e, T_f)}$$

$$\frac{17.034}{1.07} = \frac{(24.5 + 3.75X)(21.1 - 4.166X)}{(6.1207 - 0.866) \times (24.5 + 3.75X)(1.936 - 6.1207) \times (21.1 - 4.166X)}$$

$$\Rightarrow \qquad 15.92 = \frac{516.95 - 22.942X - 15.623X^2}{40.4365 + 37.1398X}$$

$$\Rightarrow \quad 15.623\ X^2 + 614.2076X + 126.7991 = 0$$

$$\Rightarrow \quad X = \log_{10}(r_e) = \frac{-614.2076 \pm \sqrt{377250.9757 - 4 \times 126.7991 \times 15.623}}{31.346}$$

$$= -\,0.206877 \text{ or } -\,38.9820$$

$$\Rightarrow \quad r_e = 0.62 \text{ or } 1.04.$$

To find the final blend ratio *R*, when *B*(*R*, *T*) = 12.52, when *T* = 6.1207

We have, $B(R, 6.1207) = 12.52$

To get $B(R, T)$, for $T = 6.1207$, we **require** [putting $Y = \log_{10} R$]

$$B(R, T_i) = C_i + m_i \log R = 24.5 + 3.75 \log R = 24.5 + 3.75\ Y$$

$$B(R, T_f) = C_f + m_f \log R = 21.1 - 4.166 \log R = 21.1 - 4.166Y$$

$$B(R, T) = \frac{B(R,T).\,B(R,T_f) \times (T_f - T_i)}{(T - T_i) \times B(R,T_i) + (T_f - T) \times B(R,T_f)}$$

$$\frac{12.52}{1.07} = \frac{(24.5 + 3.75Y)(21.1 - 4.166Y)}{(6.1207 - 0.866) \times (24.5 + 3.75Y) + (1.936 - 6.1207) \times (21.1 - 4.166Y)}$$

$$\Rightarrow \qquad 11.70 = \frac{516.95 - 22.942Y - 15.623Y^2}{40.4365 + 37.1398Y}$$

$$\Rightarrow \quad 15.623\ Y^2 + 457.478Y - 43.843 = 0$$

$$\Rightarrow \qquad Y = \log_{10} R = \frac{-\,457.478 \pm \sqrt{209286.120 + 4 \times 15.623 \times 43.843}}{31.346}$$

$$= \frac{-\,457.478 \pm 460.463}{31.346} = 0.0952 \quad \text{or,} \quad -\,29.284$$

$\Rightarrow$ Taking the first value, $R = 1.245$

Substitute now, $R = 1.245$ and $r_e = 1.04$ in

$$R = r\,(1 + r_e) + r_e \quad \Rightarrow \quad \mathbf{r = 0.1.}$$

Thus, the **proportion of the fresh slurry required to be blended with the basic partially set M20 concrete at *t* = 2.25 hr, by weight, to yield the final crossed strength of 12.52 MPa, is 10 percent.**

[**Note:** The experiment was conducted to find the final strength of the slurry mixed concrete adopting the blend ratio $r = 0.1$. The result was 12.52 MPa. The said problem is **verified** analytically here in the reverse way.]

4.5 VERIFICATION OF PLATE DEFLECTION ANALOGY

Example 4.16. *A thin rectangular plate of size (a × b) is described through non-dimensional coordinates*

$(\bar{x}, \bar{y})$, where $\bar{x} = x / a$ and $\bar{y} = 2y / b$, is simply supported along the three edges, viz., $\bar{x} = 0$, $\bar{y} = 0$ and $\bar{y} = 2$, and free along the fourth edge $\bar{y} = 1$. The plate carries a uniform loading on its surface. The deflection characteristic of this type of plate is described in **Art. 3.2 (Item 4)**, *and in* **Fig. 3.6**.

Verify *the efficiency of the approximate formula* **(3.17)** *for the deflection $w(\bar{x}, \bar{y})$ at various points on the half portion of the plate, bounded by ($\bar{x} = 0$ and $\bar{x} = 1$), and ($\bar{y} = 0$ and $\bar{y} = 1$), and compare with the corresponding deflections obtained from closed form solution using* **Levy's method**, *in reference to the maximum deflection which occurs at $w(1, 1)$.*

Solution. Refer to **Art. 3.2 (Item 4),** and **Fig. 3.6.**

The locations $(\bar{x}, \bar{y})$ at which the deflections calculated are:

For $\quad \bar{x} = \frac{1}{8}, \frac{1}{4}, \frac{3}{8}, \frac{1}{2}, \frac{5}{8}, \frac{3}{4}, \frac{7}{8}, 1$

Along $\bar{y} = \frac{1}{8}$ $\quad\Rightarrow\quad$ points 1, 2, 3, 4, 5, 6, 7, 8, respectively.

Along $\bar{y} = \frac{1}{2}$, $\quad\Rightarrow\quad$ points 9, 10, 11, 12, 13, 14, 15, 16, respectively.

Along $\bar{y} = \frac{3}{4}$, $\quad\Rightarrow\quad$ points 17, 18, 19, 20, 21, 22, 23, 24, respectively.

Along $\bar{y} = 1$, $\quad\Rightarrow\quad$ points 25, 26, 27, 28, 29, 30, 31, 32, respectively.

Along the boundary lines, $\bar{y} = 0$ (*i.e.*, $\bar{x}$ – axis), and $\bar{x} = 0$ (*i.e.*, $\bar{y}$ –axis), deflection $w = 0$.

The maximum deflection is at point 32, *i.e.*, $w_{32} = w(1,1)$. Take w_{32} = **100** (say).

All other deflections, $w_1, \ldots w_{32}$, are computed at various locations $(\bar{x}, \bar{y})$ relative to this value.

The detailed calculation of the true values obtained from the closed form analytical solution of the deflections at various points using Levy's method, are not presented here as it is, but are shown **as relative values with respect to the maximum deflection as 100, occurring at point 32.**

Side by side, all the corresponding deflections are computed at points 1, ...31, considering $w(1,1) = w_{32}$ = **100**, using the approximate formula (3.17). Thus

$$w(\bar{x}, \bar{y}) = \frac{w(1, \bar{y}) \, . \, w(\bar{x}, 1)}{w(1, 1)} = \frac{w(1, \bar{y}) \, . \, w(\bar{x}, 1)}{100}.$$

The approximate formula (3.17) works only when, out of four deflection quantities, viz., $w(\bar{x}, \bar{y})$, $w(1, \bar{y})$, $w(\bar{x}, 1)$ and $w(1,1)$, **if three are known, the fourth can be found.**

All other deflection curves along $\bar{x}$ = constant, (*i.e.*, lines parallel to $\bar{y}$-axis) are sinusoidal, with zero value at $\bar{y} = 0$, and amplitude at $\bar{y} = 1$, the trend of which is described as

$$w(\bar{x} = \text{constant}, \bar{y}) = C \sin\left(\frac{\pi . \bar{y}}{2}\right)$$

So, along the free edge, where $\bar{x} = 1$, the sinusoidal deflection curve is having maximum ordinate at (1, 1), which has been be taken as the amplitude of magnitude **(given) = 100**.

Thus, deflection ordinates $w(1, \bar{y})$ can easily be calculated using

$$w(1, \bar{y}) = 100 \sin\left(\frac{\pi . \bar{y}}{2}\right)$$

The values are : 38.268, 70.711, 92.388, and 100.00 corresponding to

$$\bar{y} = \frac{1}{4}, \frac{1}{2}, \frac{3}{4}, 1$$

The results are presented in **Table 4.1** for the inside points 1, 2, ...31, on the plate surface. The format in which the results are recorded for each deflection location $(\bar{x}, \bar{y})$ are as follows:

At each location $(\bar{x}, \bar{y})$, the identification of the point occupies top most position. Just down below is the deflection using plate solution, and at the bottom is the deflection using approximate relation (3.17).

[XXXX = deflection values on the boundary line $\bar{y} = 1$, that are not calculated., and

NNNN = deflection values on the inside points that are not calculated]

Table 4.1 Comparison of Deflection Using Plate Solution and Approximate formula (3.17)

$\bar{x} \rightarrow$ $\bar{y} \downarrow$	1/8	1/4	3/8	1/2	5/8	3/4	7/8	1
	Point 1	**Point 2**	**Point 3**	**Point 4**	**Point 5**	**Point 6**	**Point 7**	**Point 8**
1/4	6.334	18.093	27.247	32.274	35.606	36.668	37.880	38.269
	6.307	18.005	27.115	32.117	35.436	NNNN	37.696	38.268
	Point 9	**Point 10**	**Point 11**	**Point 12**	**Point 13**	**Point 14**	**Point 15**	**Point 16**
1/2	11.710	33.430	50.346	59.633	65.797	67.754	69.994	70.711
	11.653	33.223	50.102	59.344	65.478	NNNN	69.654	70.711
	Point 17	**Point 18**	**Point 19**	**Point 20**	**Point 21**	**Point 22**	**Point 23**	**Point 24**
3/4	15.300	43.680	65.780	77.914	85.967	NNNN	91.451	92.388
	15.226	43.467	65.460	77.536	85.550	NNNN	NNNN	92.388
	Point 25	**Point 26**	**Point 27**	**Point 28**	**Point 29**	**Point 30**	**Point 31**	**Point 32**
1	16.561	47.278	71.199	84.344	93.050	95.818	98.986	100.00
	XXXX	XXXX	XXXX	XXXX	XXXX	XXXX	XXXX	100.00

Note: A comparison of the results at each location shows that the approximate formula (3.17) predicts the intermediate results with a very high degree of accuracy, and the range of error of the deflection predicted by the approximate formula (3.17) with respect to that by closed form plate solution, lies around - 0.48 %.

Thus, **the approximate formula (3.17) can be safely used for all practical purpose for calculating the relative values of the deflection of the internal points with respect to the maximum value of the deflection in a loaded thin plate.**

[See next two examples (**Example 4.17, Example 4.18**) in support of this statement.]

Example 4.17. *Calculate the deflection of the plate (Example 16) at point 27, which is shown in Table 4.1 as XXXX., using approximate formula (3.17).*

Solution. Recollect that in the formula *viz.,*

$$w(\bar{x}, \bar{y}) = \frac{w(1, \bar{y}) \,.\, w(\bar{x}, 1)}{w(1, 1)} = \frac{w(1, \bar{y}) \,.\, w(\bar{x}, 1)}{100}$$

The four points $(\bar{x}, \bar{y})$, $(1, \bar{y})$, $(\bar{x}, 1)$ and (1,1) form a rectangular mesh, where points $(1, \bar{y})$ and $(\bar{x}, 1)$ are the locations at the two extreme boundary lines *viz.*, $\bar{x} = 1$, and $\bar{y} = 1$, respectively, and are diagonally opposite to each other.

The other pair of locations *viz.*, ($\bar{x}$, $\bar{y}$), and (1,1) also occupy two ends of the other diagonal of the rectangular grid.

Out of four deflection quantities, *viz.*, $w(\bar{x}, \bar{y})$, $w(1, \bar{y})$, $w(\bar{x}, 1)$ and $w(1, 1)$, if three are known, then the fourth can be found either using (3.16) or (3.17).

Here, **consider the rectangular grid as 11, 16, 32 and 27.**

Refer to the values from Table 4.1.

w_{11}**(given)** = 50.102 **(inside point)**

w_{16} **(given)** = 70.711 **(on extreme boundary** $\bar{x}$ = 1)

w_{27} **(to be determined) (on extreme boundary** $\bar{y}$ = 1)

w_{32} **(given)** = 100 **(extreme corner point)**

Apply $w(\bar{x}, \bar{y}) = \dfrac{w(1, \bar{y})\,.\,w(\bar{x}, 1)}{w(1, 1)}$

$$\Rightarrow \qquad w_{27} = \frac{w_{11}.w_{32}}{w_{16}} = \frac{50.102 \times 100}{70.711} = \mathbf{70.855\ (71.199)}$$

[The value within parenthesis is that from plate solution. **Error = – 0.48 %**].

Example 4.18. *Using data from Table 4.1 of Example 4.16, verify that the product of the deflection values occurring at the end points of each of the two diagonals of any rectangular grid on plate surface (as shown in Fig. 3.6), are equal.*

Solution. To prove that, for any rectangular grid *ABCD* of which *AC* and *BD* are the diagonals,

$w_A \,.\, w_C = w_B \,.\, w_D$, take two sets of data as follows:

1. Choose points 12, 13, 21 and 20.

$$w_A = w_{12} = 59.344,\ w_B = w_{13} = 65.478,$$

$$w_C = w_{21} = 85.550 \text{ and } w_D = w_{20} = 77.536$$

$$w_A \,.\, w_C = 59.344 \times 85.550 = \mathbf{5076.8792},$$

and $$w_B \,.\, w_D = 65.478 \times 77.536 = \mathbf{5076.9022}.$$

2. Choose points 2, 5, 21 and 18.

$$w_A = w_2 = 18.005,\ w_B = w_5 = 35.436,$$

and $$w_C = w_{21} = 85.550,\ w_D = w_{18} = 43.467$$

$$w_A \,.\, w_C = 18.005 \times 85.550 = \mathbf{1540.3278},$$

and $$w_B \,.\, w_D = 35.436 \times 43.467 = \mathbf{1540.2530}$$

Thus $w_A \,.\, w_C = w_B \,.\, w_D$, for both sets of data (**verified**).

Example 4.19. *Using differential to approximate the strength value of A(∞, 2.67) at t = 2.67, when the strength A(∞ , 2) = 145 kg/sq cm at t = 2, is given. The other data is same as in* ***Example 4.5(b).***

Solution. Recall (3.4) as

$$A_{r,t} = \frac{A_{r,i} . A_{r,f} . (t_f - t_i)}{t . (A_{r,i} - A_{r,f}) + A_{r,f} . t_f - A_{r,i} . t_i} \tag{4.10}$$

$$\frac{d}{dt}(A_{rt}) = \frac{- A_{r,i} . A_{r,f} . (A_{r,i} - A_{r,f}) . (t_f - t_i)}{[t . (A_{r,i} - A_{r,f}) + A_{r,f} . t_f - A_{r,i}\, t_i]^2} \tag{4.11}$$

Here (**given**): $t = 2, \Delta t = 0.67, A_{rt} = 145, \Delta A_{rt} =$ **(to find)**

Now $$d(A_{rt}) = \Delta A_{rt} = \frac{d}{dt}(A_{rt}).dt$$

$$= \frac{-(160) \times (105) \times (55) \times (1.25)}{[2 \times 55 + (3 \times 105 - 1.75 \times 160)]^2} \times (0.67) = \mathbf{-36.8}$$

Thus $A(\infty, 2.67) = A(\infty, 2) - 36.8 = 145 - 36.8 =$ **108.2** kg/sq cm (**111.5**)

Results differ by **3.05 %**.

CONCLUDING REMARKS

The scope of this book has been kept limited to the introduction of the Selfing and Crossing Concept, and development of the technical theory based on the Concept, and side by side, demonstration of the versatility of the proposed formulation to a wide range of actual field problems concerning partially set concretes in blending.

The Selfing formula is capable of computing analytically all strengths in the given time lag range and blend ratio range, if for each constituent mix in blending, the two strengths at the two setting times for the basic mix (for which $r = \infty$), and the suitable relations of the strengths at the two setting times only, as functions of r, (leading to characteristic strength parameters), are known.

The direct application of the Selfing formulation involving strengths as function of the blend ratio r, and time lag t as $A(r, t)$, has been demonstrated and lucidly elaborated with day to day concreting problems. However, there are few identified cases of practical field problems like, assessing of the strength at the interface of two partially set concrete layers, cast one over the other layer, at a certain time gap, and that of the blended fiber reinforced concretes, each prepared with certain dosage of steel fiber and observed at certain time lag, have also been brought under the umbrella of the said formulation by suitably modifying the formulation analogously.

This modification is necessary, since the basic Selfing equation, though appears structurally unchanged, cannot be applied directly. In the former modification, the interfacial strength, *viz.*, the bond strength $I(M, T)$ is considered as the function of the mix grade number M (like, M20, M30, etc.), and the time lag parameter T (= t^n, where n is any suitable positive number), and in the latter modification, the selfed strength of the fibre reinforced concrete $F(p, t)$ is considered as the function of steel fibre dosage p, and the time lag t.

The Selfing equation for $A(r, t)$ has been derived on the assumption that, at $t = 0$, the strength is maximum, and they droop down thereafter as the time lag progresses. However, the strength-time curves [*i.e.*, $A(r, t)$ – variation] of some concretes often demonstrate rising trend from that at $t = 0$, up to or around $t = t_i$, and then follow drooping trend as usual. In such cases, the Selfing equation can be applied without any difficulty by breaking the total time lag zone, ($t = 0$ to $t = t_f$), into two Zones, *viz.*, Zone I : by taking t_i as 0, and t_f as t (the observed time where the maximum upward trend is noticed), and Zone II : by taking t_i as t, and t_f as the given t_f, of the mix. Cases where the strength-time curves show all rising trend from that at $t = 0$, does not pose any difficulty. They can be tackled directly using the basic Selfing equation, and the corresponding characteristic strength parameters.

For measuring shear strength, the usual procedure followed is to cause slip at one or two planes during relative sliding, which technically is termed as single shear or double shear. Experience shows that for getting single shear under compression testing machine is unreliable. This is because of the fact that, too much eccentricity is caused as one portion of the specimen requires to be held fixed with respect to the other part during failure. On the other hand, failure of two planes simultaneously in double shear never happens in practice. No clear cut provision of the codes, as guidelines, is available in this regard. So altogether a new technique has been devised and successfully applied for obtaining highly accurate, reliable and uniform values of shearing strength. The specimen is an inverted *L*-shaped prism (cast inside the standard cube mould by placing a solid wooden block at one corner), is held under the compression testing machine in such a way that the axial load P from the machine, is transmitted as $P_1 = P.[L_2 /(L_1 + L_2)]$, to hold tightly the specimen as the clamping force, while the other part of P as $P_2 = P.[L_1/ (L_1 + L_2)]$, is utilized as the required shearing force to act at the intended section of the inverted *L*-specimen. The spacing L_1 and L_2 of the forces P_1 and P_2, respectively, from the line of action of P, is so positioned as the centroid of the forces P_1 and P_2 coincides with the line of action of P. The sharing of loads is achieved through a piece of steel plate of length L (= $L_1 + L_2$), placed between the ram of the testing machine and the specimen.

For the first time, the spring model of the behaviour of partially set concretes in selfing of two mixes, is proposed in Chapter 3 (Fig. 3.5). Each of the spring system, K_1

or K_2 is made up of two elastic springs namely, one being + ve and the other being –ve. If Selfing or Crossing is performed on more than two mixes, then in each set of *K*-s, the arrangement of equivalent linearly elastic springs in series, will be more than two, in which the connection of +ve and –ve springs is expected to be of unknown combination, and is not at all easy to obtain the required one. Moreover, such combination in each *K*, may not turn out to be a unique one, but will have to be selected from a number of combinations. The hardship of getting the break up of equivalent linear spring combinations in series of any given stiffness *K*, is apparent when treatment is attempted analytically.

In this direction, in Appendix, a new analytical treatment of expressing the final stiffness of a system of linear elastic springs arranged in series in terms of their individual stiffnesses, and also the break up of a given stiffness of a spring into the corresponding stiffnesses of a number of equivalent linear elastic springs, arranged in series, are discussed on developing and proposing an altogether new concept of analytical treatment, known as Spring Function. The basic Spring Function and its application is available in an earlier publication elsewhere, and further extension of the said Spring Function is elaborated in details in Appendix. The utility of this proposed Spring Function may be of help for future research workers for analytical treatment of such problems, if required.

It is felt that, a family of smooth curves showing variations, either of rising or decreasing trend, similar to the strength-time variation, is possible to be expressed and mathematically analysed through Selfing formulation, analogously. Example of such a problem may be concerning mixing of two different or same type of hot liquids, at some proportion, and observed at some identified elapsed time interval.

Example: Liquids heated to certain temperature and allowed to cool uninterrupted, show drooping trend with elapsed time, till the temperature of the enclosure is reached. Two liquids *A* and *B* of quantities Q_A and Q_B, each of which was heated to reach certain temperatures T_{A0} and T_{B0}, were allowed to cool to show temperatures T_A and T_B, after elapsed time t_A and t_B, $(t_B > t_A)$, respectively. At the instant of time *t*, (when both *A* and *B* attain t_A and t_B), respectively, both were taken together and thoroughly mixed in certain ratio $r\ (= Q_B / Q_A)$. Determine the resulting temperature *T* of the total mixture of the blended liquid conforming to the General Crossing formulation. If applicable, all the various Representations of the Selfing equation (Art 3.3) will automatically hold good.

[In this example, only those liquids not reacting chemically, are considered.]

APPENDIX

Spring Function and its Application

A.1 GENERAL

A new concept of expressing final stiffness of a system of linear elastic springs arranged in series in terms of their individual stiffnesses, and also the break up of a given stiffness of a spring into the corresponding stiffnesses of a number of equivalent linear elastic springs, arranged in series, is proposed.

Under axial pull P, if a spring shows extension d, the stiffness of the spring $K = P/d$ is defined as **positive**. On the contrary, if shortening results, instead of extension under pull, the stiffness of the spring K is defined as **negative**.

Refer to **Fig. 3.5** in connection with the **Rheological model (Spring model)**, and definition of Positive and Negative springs, we recollect that

Positive spring: When ***P*** is + ve, and ***d*** is + ve, or ***P*** is – ve, and ***d*** is – ve.

Negative spring: When ***P*** is + ve, and ***d*** is – ve, or ***P*** is + ve, and ***d*** is – ve.

In the present discussion, consider only **linear elastic springs** which are **arranged in series**. Two types of problems are visualized, *viz:* **Direct Problem** and **Reverse Problem**.

Direct Problem: Where if a number of springs of stiffnesses K_1, K_2, K_3, … , K_n is **given**, **find** the stiffness K of an equivalent single spring.

That is, find K in terms of K_1, K_2, K_3, … , K_n.

The relation is well known. For n-number of springs

$$\frac{1}{K} = \frac{1}{K_1} + \frac{1}{K_2} + \frac{1}{K_3} + ... + \frac{1}{K_n} \qquad (A.1)$$

Reverse Problem: Where if a single spring of stiffness K is **given**, and if it is intended to break it up as a combination of several springs, **find** the stiffnesses K_1, K_2, K_3, ..., K_n, of these individual springs.

The reverse problem has no such relation like (A.1) existing. This is because of the fact that the unknowns K_1, K_2, K_3, ..., K_n are many in comparison to the known which is one only (*i.e.*, K). In fact, for a certain number of springs replacing the single one, there could be an infinite number of possible combinations.

While the former problem is straightforward, the latter problem is not so. The **Spring Function concept** is introduced and demonstrated first in the former problems, and the latter problem is successfully dealt with this concept along with a novel method using **spring number**.

A.2 SPRING FUNCTION CONCEPT

Let n-number of springs are connected in series, and their stiffnesses are K_1, K_2, K_3, ..., K_n. Each of the stiffnesses, K_i is expressed as function of $f(s)$

$$K_1 = \frac{S_1(s)}{C_1} = \frac{s - a_1}{C_1} \quad \Rightarrow \quad S_1 = S_1(s) = (s - a_1)$$

$$K_2 = \frac{S_2(s)}{C_2} = \frac{s - a_2}{C_2} \quad \Rightarrow \quad S_2 = S_2(s) = (s - a_2)$$

$$K_3 = \frac{S_3(s)}{C_3} = \frac{s - a_3}{C_3} \quad \Rightarrow \quad S_3 = S_3(s) = (s - a_3)$$

... ...

$$K_n = \frac{S_n(s)}{C_n} = \frac{s - a_n}{C_n} \quad \Rightarrow \quad S_n = S_n(s) = (s - a_n) \qquad (A.2a)$$

where, $a_1, a_2, a_3, \ldots, a_n$, are **arbitrary constants**

and $C_1, C_2, C_3, \ldots, C_n$, are **spring numbers** and **constants,**

and $S_1, S_2, S_3, \ldots, S_n$, are **modified spring functions of the individual springs**

so that
$$\frac{1}{S} = \frac{1}{S_1} + \frac{1}{S_2} + \frac{1}{S_3} + \ldots + \frac{1}{S_n}$$

$$= \frac{1}{(s-a_1)} + \frac{1}{(s-a_2)} + \frac{1}{(s-a_3)} + \ldots + \frac{1}{(s-a_n)} \qquad (A.2b)$$

where S = **modified spring stiffness** and is **function of s.**

Define now **Spring function** $K_s = K_s(s)$ as a chain product of the **modified spring functions** as

$$K_s = S_1 \,.\, S_2 \,.\, S_3, \ldots, S_n = (s - a_1)(s - a_2)(s - a_3) \ldots . (s - a_n)$$

That is

$$K_s = \prod_{i=1}^{n} S_i = \prod_{i=1}^{n} (s - a_i) = (s - a_1)(s - a_2)(s - a_3) \ldots . (s - a_n) \qquad (A.3a)$$

Note: Since $a_1, a_2, a_3, \ldots, a_n$ are arbitrary, without any loss of generality, they can be assigned any numerical value. For the sake of simplicity, they may be given values

$$a_1 = 1, a_2 = 2, a_3 = 3, \ldots, a_n = n.$$

for which case

$$K_s = \prod_{i=1}^{n} S_i = \prod_{i=1}^{n} (s - i) = (s - 1)(s - 2)(s - 3) \ldots . (s - n) \qquad (A.3b)$$

If, (say) for the sake of argument, $K = K_s$, then we have the following identities:

For: 2-spring system

Use $(A.1)$, $(A.2a)$, and $(A.2b)$

$$\frac{1}{K} = \frac{1}{K_1} + \frac{1}{K_2} = \frac{C_1}{S_1} + \frac{C_2}{S_2} = \frac{C_1 S_2 + C_2 S_1}{S_1 S_2 = (K_s)}$$

$$\Rightarrow \qquad C_1 S_2 + C_2 S_1 = 1 \qquad (A.4a)$$

For: 3-spring system

Use $(A.1)$, $(A.2a)$ and $(A.2b)$

$$\frac{1}{K} = \frac{1}{K_1} + \frac{1}{K_2} + \frac{1}{K_3} = \frac{C_1}{S_1} + \frac{C_2}{S_2} + \frac{C_3}{S_3} = \frac{C_1 S_2 S_3 + C_2 S_1 S_3 + C_3 S_1 S_2}{S_1 S_2 S_3 = (K_s)}$$

$$\Rightarrow \qquad C_1 S_2 S_3 + C_2 S_1 S_2 + C_3 S_1 S_2 = 1 \qquad (A.4b)$$

For: 4-spring system

$$\frac{1}{K} = \frac{1}{K_1} + \frac{1}{K_2} + \frac{1}{K_3} + \frac{1}{K_4} = \frac{C_1}{S_1} + \frac{C_2}{S_2} + \frac{C_3}{S_3} + \frac{C_4}{S_4}$$

$$\Rightarrow \qquad C_1 S_2 S_3 S_4 + C_2 S_1 S_3 S_4 + C_3 S_1 S_2 S_4 + C_4 S_1 S_2 S_3 = 1 \qquad (A.4c)$$

and so on.

If this type of criteria hold true for a case, it suggests the equality of the two values of K and K_s.

Application of the relations of the type (A.4) will be shown later.

A.3 TO FIND THE VALUE OF SPRING NUMBERS C_i

Three methods have been dealt here for evaluating the values of **spring numbers** C_i, *viz.*,

Method 1: using **Partial Fraction Method,**

Method 2: using **Ready made formula 1,**

Method 3: using **Ready made formula 2, and**

Method 4: using **Antimatrix Table (proposed)**

1. Method 1: using **Partial Fraction Method**

For: 2-spring system ($n = 2$).Use (A.3a)

$$K_s = \prod_{i=1}^{2} S_i = \prod_{i=1}^{2} (s - a_i) = S_1S_2 = (s - a_1)(s - a_2) = s^2 - (a_1 + a_2)s + a_1a_2$$

Expand the reciprocal of K_s in partial fraction as

$$\frac{1}{K_s} = \frac{1}{(s-a_1)(s-a_2)} = \frac{C_1}{(s-a_1)} + \frac{C_2}{(s-a_2)}$$

$$= \frac{s.(C_1 + C_2) - (C_1a_2 + C_2a_1)}{(s-a_1)(s-a_2)}$$

Since the numerator = 1, we get two simultaneous equations in C_1 and C_2 as

$$C_1 + C_2 = 0$$

$$C_1a_2 + C_2a_1 = -1 \qquad (A.5a)$$

Solving for $C_i - s$

$$C_1 = \frac{1}{(a_1 - a_2)}, \text{ and } C_2 = \frac{-1}{(a_1 - a_2)}$$

or $$\frac{1}{C_1} = (a_1 - a_2), \text{ and } \frac{1}{C_2} = -(a_1 - a_2) \qquad (A.5b)$$

For the case when $a_i - s$ take the values progressively, starting with 1 as 1, 2, 3, ...n:

For: 2-spring system with $\{a_1, a_2\} = \{1, 2\}$

that is, for $K_s = S_1 S_2 = (s-1)(s-2)$ (*A.5c*)

From (*A.5b*):

$$C_1 = -1, \quad C_2 = 1$$

$$\Rightarrow \quad \frac{1}{C_1} = -1, \frac{1}{C_2} = 1$$

(*A.5d*)

For: 3-spring system ($n = 3$).Use (*A.3*).

$$K_s = \prod_{i=1}^{3} S_i = \prod_{i=1}^{3} (s - a_i) = S_1 S_2 S_3 = (s-a_1)(s-a_2))(s-a_3)$$

Expand the reciprocal of K_s in partial fraction as

$$\frac{1}{K_s} = \frac{1}{(s-a_1)(s-a_2)(s-a_3)} = \frac{C_1}{(s-a_1)} + \frac{C_2}{(s-a_2)} + \frac{C_3}{(s-a_3)}$$

$$= \frac{s^2 .(C_1 + C_2 + C_3) - s.[C_1(a_2 + a_3) + C_2(a_3 + a_1) + C_3(a_1 + a_2)] + [C_1 a_2 a_3 + C_2 a_3 a_1 + C_3 a_1 a_2]}{(s-a_1)(s-a_2)(s-a_3)}$$

Since the numerator = 1, we get three simultaneous equations in C_1, C_2 and C_3 as

$$C_1 + C_2 + C_3 = 0$$

$$C_1(a_2 + a_3) + C_2(a_3 + a_1) + C_3(a_1 + a_2) = 0$$

$$C_1 \cdot a_2\, a_3 + C_2 \cdot a_3\, a_1 + C_3 \cdot a_1\, a_2 = 1 \quad (A.5e)$$

Solving (*A.5d*), C_1, C_2 and C_3 are obtained.

For the case when $a_i - s$ take the values progressively, starting with 1 as 1, 2, 3, ...n:

For: 3-spring system with $\{a_1, a_2, a_3\} = \{1, 2, 3\}$

that is, for $K_s = S_1 S_2 S_3 = (s-1)(s-2)(s-3)$ (*A.5f*)

From (*A.5e*):

$$C_1 = \frac{1}{2}, C_2 = -1, C_3 = \frac{1}{2}$$

$$\Rightarrow \quad \frac{1}{C_1} = 2, \frac{1}{C_2} = -1, \frac{1}{C_1} = 2 \quad (A.5g)$$

For: *n*-spring system ($n = n$).Use ($A.3a$).

$$K_s = \prod_{i=1}^{n} S_i = \prod_{i=1}^{n} (s - a_i) = (s - a_1)(s - a_2)(s - a_3) \;....\; (s - a_n)$$

Expand the reciprocal of K_s in partial fraction as

$$\frac{1}{K_s} = \frac{1}{(s-a_1)(s-a_2)\ldots(s-a_n)}$$

$$= \frac{C_1}{(s-a_1)} + \frac{C_2}{(s-a_2)} + \frac{C_3}{(s-a_3)} + \ldots + \frac{C_n}{(s-a_n)}$$

$$= \frac{\text{Numerator}}{(s-a_1)(s-a_2)\ldots(s-a_n)}$$

The numerator of the RHS is a polynomial in s of degree $(n - 1)$.

In matrix notation, the n-number of simultaneous equations with $C_i - s$ as unknowns, is written as

$$[A]_{n \times n} \{C\}_{n \times 1} = \{N\}_{n \times 1} \tag{A.6}$$

The square matrix $[A]$ of size $(n \times n)$ is of elements written with $a_i - s$. The column matrix $\{C\}$ of the unknowns comprising of the constants $C_1, C_2, C_3, \ldots, C_n$, is of size $(n \times 1)$, corresponding to $s^{n-1}, s^{n-2}, \ldots, s^0$. The RHS column matrix $\{N\}$ is having the last (*i.e.*, bottom most) value of $(-1)^{n-1}$, which is either (+1) or (– 1), whereas all other values are zeros.

It can be noted that, as the number of springs increases, the tediousness of forming the simultaneous equations, and solving for the spring number $C_i - s$ also increase.

Once the spring numbers $C_i - s$ are available in hand, the spring stiffness $K_i - s$ can now be expressed through the modified spring stiffness $S_i - s$, as function of s, so that the real spring system involving several springs in series can be expressed through the spring function K_s using ($A.3$) in the light of ($A.2a$).

2. Method 2: using **Ready Made Formula 1**

The tediousness of forming the simultaneous equations, and solving for the spring numbers C_i is evident in Method 1, can be simply avoided if the following formula is made use of.

$$C_i = \left(\frac{s - a_i}{K_s}\right)_{s=a_i} \tag{A.7a}$$

where K_s is given by ($A.3a$).

Check

For: 2-spring system ($n = 2$).Use ($A.3a$).

$$C_1 = \left(\frac{(s-a_1)}{(s-a_1)(s-a_2)} \right)_{s=a_1} = \frac{1}{(a_1 - a_2)}$$

$$C_2 = \left(\frac{(s-a_2)}{(s-a_1)(s-a_2)} \right)_{s=a_2} = \frac{1}{(a_2 - a_1)} = -\frac{1}{(a_1 - a_2)} \qquad (A.7b)$$

For: 2-spring system with $\{a_1, a_2\} = \{1, 2\}$. Use ($A.7b$)

$C_1 = -1$, $C_2 = 1$

For 3-spring system ($n = 3$).Use ($A.3$).

$$C_1 = \left(\frac{(s-a_1)}{(s-a_1)(s-a_2)(s-a_3)} \right)_{s=a_1} = \frac{1}{(a_1 - a_2)(a_1 - a_3)}$$

$$C_2 = \left(\frac{(s-a_2)}{(s-a_1)(s-a_2)(s-a_3)} \right)_{s=a_2} = \frac{1}{(a_2 - a_1)(a_2 - a_3)} \qquad (A.7c)$$

$$C_3 = \left(\frac{(s-a_3)}{(s-a_1)(s-a_2)(s-a_3)} \right)_{s=a_3} = \frac{1}{(a_3 - a_1)(a_3 - a_2)} \qquad (A.7d)$$

For: 3-spring system with $\{a_1, a_2, a_3\} = \{1, 2, 3\}$. Use ($A.7c$).

$$C_1 = \left(\frac{(s-1)}{(s-1)(s-2)(s-3)} \right)_{s=1} = \frac{1}{(-1)(-2)} = +\frac{1}{2}$$

$$C_2 = \left(\frac{(s-2)}{(s-1)(s-2)(s-3)} \right)_{s=2} = \frac{1}{(1)(-1)} = -1$$

$$C_3 = \left(\frac{(s-3)}{(s-1)(s-2)(s-3)} \right)_{s=3} = \frac{1}{(2)(1)} = +\frac{1}{2}$$

For: 4-spring system with $\{a_1, a_2, a_3, a_4\} = \{1, 2, 3, 4\}$.

$K_s = S_1 S_2 S_3 S_4 = (s-1)(s-2)(s-3)(s-4)$

Recall Partial Fraction Method (Method 1):

$$\frac{1}{(s-a_1)(s-a_2)(s-a_3)(s-a_4)} = \frac{C_1}{(s-a_1)} + \frac{C_2}{(s-a_2)} + \frac{C_3}{(s-a_3)} + \frac{C_4}{(s-a_4)}$$

$$\Rightarrow \qquad \frac{1}{(s-1)(s-2)(s-3)(s-4)} = \frac{C_1}{(s-1)} + \frac{C_2}{(s-2)} + \frac{C_3}{(s-3)} + \frac{C_4}{(s-4)}$$

Use (*A.7a*) with $\{a_1, a_2, a_3, a_4\} = \{1, 2, 3, 4\}$.

$$C_1 = \frac{1}{(-1)(-2)(-3)} = -\frac{1}{6}$$

$$C_2 = \frac{1}{(1)(-1)(-3)} = +\frac{1}{2}$$

$$C_3 = \frac{1}{(1)(-1)(-2)} = -\frac{1}{2}$$

$$C_4 = \frac{1}{(3)(2)(1)} = +\frac{1}{6} \qquad (A.7e)$$

The $\boldsymbol{B_i = 1 / C_i}$ values are recorded in **Table A.1**.

Table A.1 Spring Number Coefficients, $B_i = 1 / C_i$

$n \rightarrow$ $j \downarrow$	$n = 2$	$n = 3$	$n = 4$	$n = 5$	$n = 6$	$\rightarrow$
$j = 1$	$B_1 = -1$	$B_1 = +2$	$B_1 = -6$	$B_1 = +24$	$B_1 = -120$	$\rightarrow$
$j = 2$	$B_2 = +1$	$B_2 = -1$	$B_2 = +2$	$B_2 = -6$	$B_2 = +24$	$\rightarrow$
$j = 3$		$B_3 = +2$	$B_3 = -2$	$B_3 = +4$	$B_3 = -12$	$\rightarrow$
$j = 4$			$B_4 = +6$	$B_4 = -6$	$B_4 = +12$	$\rightarrow$
$j = 5$				$B_5 = +24$	$B_5 = -24$	$\rightarrow$
$j = 6$					$B_6 = +120$	$\rightarrow$
$\downarrow$	$\downarrow$	$\downarrow$	$\downarrow$	$\downarrow$	$\downarrow$	$\rightarrow$ $\downarrow$

3. Method 3: using **Ready Made Formula 2**

The pattern of the spring numbers can be described through another simple formula valid for $K_s = S_1 S_2 S_3 S_4 \ldots S_n = (s-1)(s-2)(s-3)(s-4)\ldots(s-n)$, with $\{a_1, a_2, a_3, a_4, \ldots a_n\} = \{1, 2, 3, 4, \ldots n\}$,which is suggested here, with $(\,)_F$ as the **factorial sign**, and remembering that $(\mathbf{0})_F = 1$, as

$$\frac{1}{C_j} = B_j = (-1)^{n-j} \cdot (j-1)_F \cdot (n-j)_F \,;\, j = 1 \text{ to } n.$$

Thus

For: 2-spring system ($n = 2$)

$$j = 1,\ \frac{1}{C_1} = B_1 = (-1)^{2-1} \cdot (1-1)_F \cdot (2-1)_F = -1$$

$$j = 2,\ \frac{1}{C_2} = B_2 = (-1)^{2-2} \cdot (2-1)_F \cdot (2-2)_F = +1$$

For: 3-spring system ($n = 3$)

$$j = 1,\ \frac{1}{C_1} = B_1 = (-1)^{3-1} \cdot (1-1)_F \cdot (3-1)_F = +2$$

$$j = 2,\ \frac{1}{C_2} = B_2 = (-1)^{3-2} \cdot (2-1)_F \cdot (3-2)_F = -1$$

$$j = 3,\ \frac{1}{C_3} = B_3 = (-1)^{3-3} \cdot (3-1)_F \cdot (3-3)_F = +2$$

For: 4-spring system ($n = 4$)

$$j = 1,\ \frac{1}{C_1} = B_1 = (-1)^{4-1} \cdot (1-1)_F \cdot (4-1)_F = -6$$

$$j = 2,\ \frac{1}{C_2} = B_2 = (-1)^{4-2} \cdot (2-1)_F \cdot (4-2)_F = +2$$

$$j = 3,\ \frac{1}{C_3} = B_3 = (-1)^{4-3} \cdot (3-1)_F \cdot (4-3)_F = -2$$

$$j = 4,\ \frac{1}{C_4} = B_4 = (-1)^{4-4} \,.\, (4-1)_F \,.\, (4-4)_F = +6$$

For: 5-spring system ($n = 5$)

$j = 1, 2, 3, 4, 5,$

$B_1 = +24, B_2 = -6, B_3 = +4, B_4 = -6, B_5 = +24$, respectively

For: 6-spring system ($n = 6$)

$j = 1, 2, 3, 4, 5, 6$

$B_1 = -120, B_2 = +24, B_3 = -12, B_4 = +12, B_5 = -24,$

$B_6 = +\ 120$, respectively and so on.

These are recorded in **Table A.1**.

4. Method 4: using **Antimatrix Table (proposed)**

The evaluation of the numerical values of the spring numbers C_i can be performed with utmost ease from the chain multiplication of the numbers row-wise, arranged in suitable arithmetic progression series. This pattern is presented in an open ended Table form (**Table A.2**), to accommodate the given value of n along horizontal rows (extending right side wise), and the total number of the corresponding $1/C_i$ values along the vertical columns (extending downwards).

The constants $a_1, a_2, a_3, a_4, \ldots$, appearing in the spring constants $K_i = (s - a_i)$, are arbitrary, so without any loss of generality, they have been taken as 1, 2, 3, 4, …, and with these, **Table A.2** is designed for $K_s = S_1S_2S_3S_4 \ldots S_n = (s-1)(s-2)(s-3)(s-4) \ldots. (s-n)$, in its simplest form.

This Table is an **Antimatrix form**, since for any square size $(n \times n)$:

(1) The leading diagonal is made with number **0**,(value wise, it is taken as: $[\mathbf{0} \Rightarrow$ factorial $\mathbf{0} = (\mathbf{0})_F = 1]$

(2) Numbers in the rows; (1 to n), in the form of arithmetical progression series, follow progressively in decreasing order, in steps of 1, as:

Row 1: starting with 0 at (0, 1)

Row 2: starting with 1 at (0, 2)

Row 3: starting with 2 at (0, 3)

… … … … … … … … …

Row 5: starting with 4 at (0, 5)

Row n: starting with $(n - 1)$ at $(0, n)$

so that

(3) Numbers in the rows; (1 to n), in the form of arithmetical progression series, follow progressively in increasing order, in steps of 1, as:

Column1: starting with 0 at (1, 0)

Column 2: starting with – 1 at (2, 0)

Column 3: starting with – 2 at (3, 0)

… … … … … … … … … …

Column 5: starting with – 4 at (5, 0),

Column n: starting with $-(n - 1)$ at $(n, 0)$

and so on.

(4) For any square matrix, $a_{kl} = -a_{lk}$. This is evident from the fact that the upper triangular matrix of the arrangement of numbers is negative with respect to

those of the lower triangular matrix, and they are separated by the diagonal made of zeros.

Table A.2 Spring Number Table

$\frac{1}{C_i}$	.... $n = 5$ .... $n = 4$ .... $n = 3$ .... $n = 2$						
$\frac{1}{C_1}$	**0**	– 1	– 2	– 3	– 4	– 5	→
$\frac{1}{C_2}$	1	**0**	– 1	– 2	– 3	– 4	→
$\frac{1}{C_3}$	2	1	**0**	– 1	– 2	– 3	→
$\frac{1}{C_4}$	3	2	1	**0**	– 1	– 2	→
$\frac{1}{C_5}$	4	3	2	1	**0**	– 1	→
$\frac{1}{C_6}$	5	4	3	2	1	**0**	→
↓	↓	↓	↓	↓	↓	↓	→

To get the required values of the spring numbers C_i using Table *A*.2, simply multiply the numbers of the horizontal rows up to the limit for the given value of *n*. For this, at times, the number **0** comes into picture. To avoid the multiplication by **0**, consider the said **0** ⇒ as factorial **0**, since $(\mathbf{0})_F = 1$. Alternatively, the said diagonal element **0** can be simply taken as absent.

Then with **0** ⇒ $(\mathbf{0})_F = 1$,

For: 2-spring system ($n = 2$): Limit is, 2 rows and 2 columns.

$$\frac{1}{C_1} = (\mathbf{0})\,(-1) = -1,\ \frac{1}{C_2} = (1)\,(\mathbf{0}) = +1$$

For: 3-spring system ($n = 3$): Limit is, 3 rows and 3 columns.

$$\frac{1}{C_1} = (\mathbf{0})\,(-1)\,(-2) = +2,\ \frac{1}{C_2} = (1)\,(\mathbf{0}) = +1,\ \frac{1}{C_3} = (2)\,(1)\,(\mathbf{0}) = +2$$

For: 4-spring system (n = 4): Limit is, 4 rows and 4 columns.

$$\frac{1}{C_1} = (\mathbf{0})\,(-1)\,(-2)\,(-3) = -6,\ \frac{1}{C_2} = (1)\,(\mathbf{0})\,(-1)\,(-2) = +2$$

$$\frac{1}{C_3} = (2)\,(1)\,(\mathbf{0})\,(-1) = -2,\ \frac{1}{C_4} = (3)\,(2)\,(1)\,(\mathbf{0})\,(-1) = +6$$

... ...

For: 6-spring system (n = 6): Limit is, 6 rows and 6 columns.

$$\frac{1}{C_1} = (\mathbf{0})(-1)(-2)(-3)(-4)(-5) = -120,\ \frac{1}{C_2} = (1)(\mathbf{0})(-1)(-2)(-3)(-4)$$

$$= +24$$

$$\frac{1}{C_3} = (2)(1)(\mathbf{0})(-1)(-2)(-3) = -12,\ \frac{1}{C_4} = (3)(2)(1)(\mathbf{0})(-1)(-2) = +12$$

$$\frac{1}{C_5} = (4)(3)(2)(1)(\mathbf{0})(-1) = -24,\quad \frac{1}{C_6} = (5)(4)(3)(2)(1)(\mathbf{0}) = +120\,.$$

A.4 RELATION BETWEEN SPRING FUNCTION AND ITS FIRST DERIVATIVE

In **Art. A.2**, the spring function concept has been elaborated. The usefulness of the spring function K_s is well-suited in the **Reverse Problem** when for a given single spring of stiffness K, it is intended to find the equivalent number of springs connected in series along with their individual stiffness $K_i - s$.

To clarify the Reverse Problem recapitulate further:

the stiffness of a single spring K is **given**, and if it is split up of:

2 number of springs in series, **what will be** the corresponding K_1, K_2 ?

3 number of springs in series, **what will be** the corresponding K_1, K_2, K_3 ?

4 number of springs in series, **what will be** the corresponding K_1, K_2, K_3, K_4 ?

n number of springs in series, **what will be** the corresponding $K_1, K_2, K_3, \ldots K_n$?

and so on.

Consider the **Direct Problem**. First, let us establish an important relation connecting the spring function K_s and its first derivative K_s'.

For: 2-spring system (n = 2)

$$K_s = S_1\,S_2 = (s - a_1)(s - a_2)$$
$$K_s' = (s - a_1) + (s - a_2) = S_1 + S_2$$

Now, in the light of ($A.2b$)

$$\frac{K_s'}{K_s} = \frac{S_1 + S_2}{S_1 S_2} = \frac{1}{S_1} + \frac{1}{S_2} = \frac{1}{S} \Rightarrow \frac{K_s}{K_s'} = S \qquad (A.8a)$$

For: 3-spring system ($n = 3$)

$$K_s = S_1\, S_2\, S_3 = (s - a_1)(s - a_2)\,)(s - a_3)$$

$$K_s' = (s - a_1)(s - a_2) + (s - a_2)(s - a_3) + (s - a_3)(s - a_1)$$

$$= S_1\, S_2 + S_2\, S_3 + S_3\, S_1$$

Now, in the light of (A.2b)

$$\frac{K_s'}{K_s} = \frac{S_1S_2 + S_2S_3 + S_3S_1}{S_1S_2S_3}$$

$$\frac{1}{S_1} + \frac{1}{S_2} + \frac{1}{S_3} = \frac{1}{S} \Rightarrow \frac{K_s}{K_s'} = S \qquad (A.8b)$$

For: 4-spring system ($n = 4$)

$$K_s = S_1\, S_2\, S_3\, S_4 = (s - a_1)(s - a_2)(s - a_3)(s - a_4)$$

$$K_s' = (s - a_1)(s - a_2)(s - a_3) + (s - a_1)(s - a_2)(s - a_4) +$$

$$(s - a_1)(s - a_3)(s - a_4) + (s - a_2)(s - a_3)(s - a_4)$$

$$= S_1\, S_2\, S_3 + S_1\, S_2\, S_4 + S_1\, S_3\, S_4 + S_2\, S_3\, S_4$$

Now, in the light of ($A.2b$)

$$\frac{K_s'}{K_s} = \frac{S_1S_2S_3 + S_1S_2S_4 + S_1S_3S_4 + S_2S_3S_4}{S_1S_2S_3S_4}$$

$$= \frac{1}{S_1} + \frac{1}{S_2} + \frac{1}{S_3} + \frac{1}{S_4} = \frac{1}{S} \Rightarrow \frac{K_s}{K_s'} = S \qquad (A.8c)$$

and so on.

Thus the **ratio of the spring function K_s and its first derivative K_s', is the modified spring stiffness S.**

The identity relation ($A.8a$) or ($A.8b$) or ($A.8c$) relating the modified stiffness constant S with the ratio of the spring function K_s and its first derivative K_s' (all being functions of s), is always true, and does not depend on the value of s or a_i.

It can be demonstrated with 3 sets of values for a 3-spring system, *viz.*,

Set 1:	$s = -7$,	$a_1 = 5$,	$a_2 = 6$,	$a_3 = -1$
Set 2:	$s = -18$,	$a_1 = 5$,	$a_2 = 6$,	$a_3 = 8$
Set 3:	$s = -10$,	$a_1 = 2$,	$a_2 = 3$,	$a_3 = 4$.

For Set 1

$$K_s = (s - a_1)(s - a_2)(s - a_3) = (-12)(-13)(-6) = -936$$

$$K_s' = (-12 \times -13) + (-13 \times -6) + (-6 \times -13) = +306$$

$$K_s/K_s' = -936/306 = -3.0588$$

$$\frac{1}{S} = \frac{1}{S_1} + \frac{1}{S_2} + \frac{1}{S_3} = -\left(\frac{1}{12} + \frac{1}{13} + \frac{1}{6}\right) = -0.3269$$

$$S = -3.0588$$

Thus $$\frac{K_s}{K_s'} = S$$

For Set 2

$$K_s = (s - a_1)(s - a_2)(s - a_3) = (-23)(-24)(-26) = -14352$$

$$K_s' = (-23 \times -24) + (-24 \times -26) + (-26 \times -23) = +1774$$

$$K_s/K_s' = -14352/1774 = -8.0902$$

$$\frac{1}{S} = \frac{1}{S_1} + \frac{1}{S_2} + \frac{1}{S_3} = -\left(\frac{1}{23} + \frac{1}{24} + \frac{1}{26}\right) = -0.12361$$

$$S = -8.0902$$

Alternatively,

$$\begin{aligned} K_s &= S_1 \cdot S_2 \cdot S_3 = (s - a_1)(s - a_2)(s - a_3) \\ &= s^3 - (a_1 + a_2 + a_3)\, s^2 + (a_1a_2 + a_2a_3 + a_3a_1)\, s - a_1a_2a_3 \\ &= -14352 \end{aligned}$$

$$\begin{aligned} K_s' &= 3\, s^2 - 2\,(a_1 + a_2 + a_3)\, s + (a_1a_2 + a_2a_3 + a_3a_1) \\ &= 3\,(-18)^2 - 2\,(19)(-18) + (30 + 48 + 40) = 1774 \end{aligned}$$

$$K_s/K_s' = -14352/1774 = -8.0902$$

Thus $$\frac{K_s}{K_s'} = S$$

For Set 3

$$K_s = (s - a_1)(s - a_2)(s - a_3) = (-12)(-13)(-14) = -2184$$

$$K_s' = (-12 \times -13)\ (-13 \times -14)\ (-14 \times -12) = +506$$

$$K_s/K_s' = -2184/506 = -4.3162$$

$$\frac{1}{S} = \frac{1}{S_1} + \frac{1}{S_2} + \frac{1}{S_3} = -\left(\frac{1}{12} + \frac{1}{13} + \frac{1}{14}\right) = -.23168$$

$$S = -4.3162$$

Thus $$\frac{K_s}{K'_s} = S$$

The said statement (A.8) is **verified.**

A.5 SOLUTION FOR REVERSE PROBLEM

In **Art. A.1** the statement of the Reverse problem has been given. Further clarification is given in **Art. A.3.** The Reverse Problem when recalled:

Given is the spring stiffness $\boldsymbol{K}$ of a single spring, and if it is intended to split it into a number of springs connected in series, **find** the individual stiffnesses, $\boldsymbol{K_1}$, $\boldsymbol{K_2}$, $\boldsymbol{K_3}$, ... $\boldsymbol{K_n}$, of the arrangement thus obtained.

The problem can best be solved by **trial and error procedure** with the help of the **spring numbers** $\boldsymbol{C_i}$, and **Spring Function** $\boldsymbol{K_s}$.

Example 1. *Given $K = 6.34$, Find K_1, K_2, K_3 and K_4 for a 4-number of spring system.*

Solution. Let us **start** with the relation $K_s = 6.34 = S_1\, S_2\, S_3\, S_4 = (s-1)(s-2)(s-3)(s-4)$, and not with the given value of $K = 6.34$. Take $S_2 = S_1 - 1$, $S_3 = S_2 - 1$, $S_4 = S_3 - 1$.

[**Note:** The reason is embedded in the criterion (A.4*c*), where it has been shown in that, the equality of the two values of K and K_s is ensured when:

$C_1S_2\, S_3\, S_4 + C_2S_1S_3\, S_4 + C_3S_1S_2\, S_4 + C_4S_1S_2\, S_3 = 1$. This criterion will be checked latter, before accepting the result.].

So, **start** with K_s as chain product of four consecutive numbers, in steps of (– 1).

Find each number, as S_1, S_2, S_3 and S_4, by trial and error method satisfying $S_2 = S_1 - 1$, $S_3 = S_2 - 1$, $S_4 = S_3 - 1$.

Do again the chain product of these numbers. It is = K_s, but now **take** it = K, [subject to verification through criterion (*A*.4*c*), which when recalled].

$$C_1S_2\, S_3\, S_4 + C_2S_1S_3\, S_4 + C_3S_1S_2\, S_4 + C_4S_1S_2\, S_3 = 1 \qquad (A.8d)$$

Having $\frac{1}{C_1}, \frac{1}{C_2}, \frac{1}{C_3}$ and $\frac{1}{C_4}$ values in hand, either from Table or from given relations, **calculate** K_1, K_2, K_3 and K_4 by multiplying the corresponding S_i and $\frac{1}{C_i}$ values, and **get** the final values of the required K_1, K_2, K_3 and K_4.

Results are recorded in **Table A.3.**

Table A.3 Trial and Error Method for the Reverse Problem (Problem 1)

S_i	S_1	S_2	S_3	S_4	Remark
Try 1 **(start)**	3.5	2.5	1.5	0.5	Π = 6.56
Try 2	3.4	2.4	1.4	0.4	Π = 4.56
Try 3	3.48	2.48	1.48	0.48	Π = 6.13
Try 4	3.45	2.45	1.45	0.45	Π = 5.51
Try 5 **(stop)**	**3.49**	**2.49**	**1.49**	**0.49**	**Π = 6.34**
$1/C_i$ = (we have)	**– 6**	**+ 2**	**– 2**	**+ 6**	From Table A.2
K_i = **(required)**	**– 20.94**	**+ 4.98**	**– 2.98**	**+ 2.94**	**Answer**

Checks

$1/ K_i = -0.0478 + 0.2008 - 0.3356 + 0.3401 = +0.1575$

⇒ from which **K** = S$(1/K_i)$ = 1/0.1575 = **6.349**

$K_s' = S_1 S_2 S_3 + S_1 S_2 S_4 + S_1 S_3 S_4 + S_2 S_3 S_4$

$= 12.948 + 1.818 + 2.548 + 4.258 = 21.572$

⇒ **S** = K_s/K_s' = 6.34/21.572 = **+ 0.294**

Again $1/S_i = 0.2865 + 0.4016 + 0.6711 + 2.0408 =$ **+ 3.4** ⇒ **S = +0.294**

To verify the criterion (*A.8d*):

LHS = – 0.3030 + 1.2740 – 2.1291 + 2.1580 = + 1 **Satisfied**

Example 2. *Show that for a given K, when split into equivalent number of K_i – s, there may exist more than one set.*

Solution. Let us start with known spring values as

Set 1: $K_1 = -12$, $K_2 = +2$, $K_3 = -10$ and $K_4 = +18$, and find equivalent single stiffness K.

$$\frac{1}{K} = \frac{1}{K_1} + \frac{1}{K_2} + \frac{1}{K_3} + \frac{1}{K_4} = -0.0833 + 0.5 - 0.1 + 0.0556 + 0.3723$$

$\Rightarrow$ $\mathbf{K = 2.687.}$

Now take the given value of $K = 2.687$ as that for a single spring, and find the four spring constants equivalent to it.

Follow exactly same procedure as of Example 1 (**Table *A*.4**)

Table A.4 Trial and Error Method for the Reverse Problem (Example 2)

S_i	S_1	S_2	S_3	S_4	Remark
Try 1 **(start)**	4	3	2	1	$\Pi = 24$
Try 2	3.5	2.5	1.5	0.5	$\Pi = 6.5625$
Try 3	3.3	2.3	1.3	0.3	$\Pi = 6.9601$
Try 4 **(stop)**	**3.28**	**2.28**	**1.28**	**0.28**	Π **= 2.6802**
$1/C_i$ = **(we have)**	**– 6**	**+ 2**	**– 2**	**+ 6**	From Table A.2
K_i = **(required)**	**– 19.68**	**+ 4.58**	**– 2.56**	**+ 1.68**	**Answer**

It can be seen that $\mathbf{K = 2.687}$ has two sets of results, namely

Set 1: $K_1 = -12$, $K_2 = +2$, $K_3 = -10$ and $K_4 = +18$, and

Set 2: $K_1 = -19.68$, $K_2 = +4.58$, $K_3 = -2.56$ and $K_4 = +1.68$.

Each set satisfies their corresponding checks.

Example 3. *Given $K = 6.34$, obtain the equivalent 4-spring, 3-spring, and 2-spring constants $K_i - s$, with $a_1 = 1$, $a_2 = 2$, $a_3 = 3$, and $a_4 = 4$, as the case may be.*

Solution. The solution through Reverse problem is same as in Example 1. The result is recorded in **Table A.5**, with $B_i = 1/C_i$. **Table A.5** shows the trial and error scheme for obtaining the required stiffnesses $K_i - s$ for a 4-, 3-, and 2-spring system.

Table A.5 Comparison of Stiffnesses for 4-, 3-, and 2-spring Systems

4-spring	$S_1 = 3.49$	$S_2 = 2.49$	$S_3 = 1.49$	$S_4 = 0.49$	S = 0.294
do	$K_1 = -20.94$	$K_2 = 4.98$	$K_3 = -2.98$	$K_4 = 2.94$	
do	$B_1 = -6$	$B_2 = 2$	$B_3 = -2$	$B_4 = +6$	
3-spring	$S_1 = 3.03$	$S_2 = 2.03$	$S_3 = 1.03$		S = 0.5575
do	$K_1 = 6.06$	$K_2 = -2.03$	$K_3 = 2.06$		
do	$B_1 = 2$	$B_2 = -1$	$B_3 = 2$		
2-spring	$S_1 = 3.065$	$S_2 = 2.065$			S = 1.234
do	$K_1 = -3.065$	$K_2 = 2.065$			
do	$B_1 = -1$	$B_2 = 1$			

Note: It is important to note that, if instead of $a_1 = 1$, $a_2 = 2$, $a_3 = 3$, we use some other arbitrary values, the values of the split for *K* for a 3-spring system, turn out to be different from those obtained with the former values of $a_i - s$.

This is done in Example 4.

Example 4. ***Given*** *K = 6.34,* ***obtain*** *the corresponding $C_i - s$ and the equivalent 3-spring constants $K_i - s$, with $a_1 = 2$, $a_2 = 5$, $a_3 = 3$,.*

Solution. For this case, Table *A*.6 shows the trial and error scheme for obtaining the required stiffnesses for a 3-spring system (exactly similar to Examples to 1 and 2), with the given value of *K* = 6.34, and with $a_1 = 2$, $a_2 = 5$, $a_3 = 3$.

Taking initially $K = 6.34 = K_s$

$$K_s = (s-2)(s-5)(s-3) = s^3 - 10s^2 + 31s - 30$$

$$K'_s = 3s^2 - 20s + 31$$

The constants (Spring Numbers), C_1, C_2 and C_3 will be different and cannot be given by (2, – 1, 2), since a_1, a_2 and a_3 are not 1, 2 and 3 respectively.

To find C_1, C_2 and C_3, call (*A*.5*e*).

The three simultaneous equations to be solved are:

$$C_1 + C_2 + C_3 = 0$$

$$8C_1 + 5C_2 + 7C_3 = 0$$

$$15C_1 + 6C_2 + 10C_3 = 1$$

The three simultaneous equations when solved yield:

$C_1 = +\,1/3,\ C_2 = +\,1/6,\ C_3 = -\,1\,/2$, from which, $B_i = 1/C_i$ values are:

$B_1 = +\,3,\ B_2 = +\,6,\ B_3 = -\,2$

These values of $B_i = 1/C_i$ are used in **Table A.6**.

Checks

$$1/\,K_i = -\,0.0912 + 0.2550 - 0.1884 = +\,0.1578$$

$\Rightarrow$ from which $\mathbf{K = 1/\Sigma(1/K_i)} = 1/0.1578 = \mathbf{6.335}$

$$1/S_i = -\,0.2737 + 0.1.5302 + 0.3769 = +\,2.1808$$

from which

$$\mathbf{S = 1/\Sigma(1/S_i)} = 1/2.1808 = \mathbf{0.4585}$$

To get the value of s, we use

$$S_1 + S_2 + S_3 = (s-2) + (s-5) + (s-3)$$

$\Rightarrow$ $6.9605 = 3\text{ s} - 10 = 6.9605 \quad \Rightarrow \quad \mathbf{s = 5.6535}$

$$K_s = (s-2)\,(s-5)\,(s-3) = s^3 - 10\,s^2 + 31s - 30$$

$$= (5.6535 - 2)\,(5.6535 - 5)\,(5.6535 - 3)$$

$$= 3.6535 \times 0.6535 \times 2.6535 = \mathbf{6.34}$$

[[or alternatively

$$K_s = [s^3 - 10\,s^2 + 31\,s - 30]_{s\,=\,5.6535} = \mathbf{6.34}]]$$

$$K'_s = [3\,\text{s}^2 - 20s + 31]_{s\,=\,5.6535} \quad \Rightarrow \quad K'_s = \mathbf{13.8092}$$

[[or alternatively

$$K'_s = S_1\,S_2 + S_2\,S_3 + S_3\,S_4$$

$$= 3.6535 \times 0.6535 + 0.6535 \times 2.6535 + 2.6535 \times 3.6535 = \mathbf{13.8092}]]$$

$S = K_s\,/K'_s = 6.335\,/13.8092 = \mathbf{0.4587.}$ **Checked**

Table A.6 Trial and Error Method for the Reverse Problem (Example 3)

S_i	S_1	S_2	S_3	Remark
Try 1 **(start)**	3.6	0.65	2.65	$\Pi = 6.287$
Try 2	3.652	0.652	2.652	$\Pi = 6.315$
Try 3	3.653	0.653	2.653	$\Pi = 6.333$
Try 4 **(stop)**	**3.6535**	**.6535**	**2.6535**	Π **= 6.335**
$1/C_i =$ **(we have)**	**+ 3**	**+ 6**	**– 2**	
$K_i =$ **(required)**	**+ 10.9605**	**+ 3.921**	**– 5.307**	**Answer**

Selected References

(ON SELFING AND CROSSING, CLOSELY RELATED, AND OTHER CITED WORKS)

Ph.D. Thesis (supervised and submitted at IIT, Bombay)

1. **Some Studies on the Strength of Blended Preset Concrete using Selfing and Crossing Concepts**

   — Goyal A.S. (1990).

2. **Some Studies on Recycled Aggregate Concrete**

   — Kishore Ravande (1994).

3. **Some Studies on the Strength of Spread Concrete using Selfing Concept**

   — Mahmoud Mohammed M. El-Fransawy (1995).

4. **Study on the use of shells as bridge decks**

   — Mohammed Ismail G. (1995).

5. **Some Studies on the Behaviour of Structural Concrete under Thermal Loading**

   — Dubal Nitin S. (1997).

6. **Some Studies on Partially Set Fibre Reinforced Concrete Under Sustained Temperature Cycle using Selfing Concept**

   — Modhera C.D. (2001).

M. Tech. Thesis (supervised and submitted at IIT, Bombay)

1. **Investigation into the Applicability of Circular Cylindrical Shells as Bridge Decks**

   — G.H. Basavarajappa (1978).

2. **Study of the Posttwisting System of Reinforcement for Concrete Beams under Torsion**

   — P.Pise (1981).

3. **Use of Circular Cylindrical Shells for Bridge Decks**

   — B.J.Gupta (1981).

4. **Study of Composite Concrete for Structural Use**

   — P.A.Joshi (1984).

5. **Study on the Use of Circular Cylindrical Shells as Bridge Decks**
   — N.V. Damodaran (1984).
6. **Study on the Effect of Cumulative Curing on the Strength of Preset Concretes**
   — M.R.Rao (1985).
7. **Study of the Strength Behaviour of Composite Mixes with Superplastisizers using Selfing Theory**
   — Gopi Siddappa (1988).
8. **Determination of the Strength of Slurry Mixed Concrete Using Crossing Theory.**
   — V.N.Ambade (1988).
9. **Study on the Behaviour of the Interface Layer between Two Preset Mixes due to Improper Casting**
   — N.Suresh (1989).

10 **Strength Behaviour of Layred Concretes**
   — M. Rame Gowda (1992).

11. **Studies on the Behaviour of Concrete Using Recycled Fine Aggregates**
    — P.Kumar (1992).
12. **Efficiency of Structural Elements with Brickwork**
    —Satish S. Hiregaudar (1992).
13. **Strength of Concrete under Cumulative Curing Condition**
    — Jagdish Sukhadeve (1995).
14. **Interface Behaviour of Layered Concretes**
    — Sanjay Wakchaure (1996).
15. **Some Studies of Concrete Beams Cast in Ferrocement Shells**
    — Parveen Kumar (2001).
16. **Behaviour of Partially Set Concretes under Sustained Temperatures**
    — Jacob G. Podipara (2002).
17. **Effect of Intermittant Curing on the Behaviour of Plain and Steel Fibre Reinforced Concrete**
    — Abhoy Kumar Mishra (2002).
18. **Strength of Composite Mix under Intermittant Curing using Selfing Theory**
    — Somnath Mishra (2003).

B.Tech. Thesis (supervised and submitted at IIT, Bombay)

1. **Behaviour of Preset Cement Mortar**
   — P.P.Jhaveri (1976).
2. **Application of Cylindrical Shells in Bridge Deck Design**
   — R.B.Bhatkar (1976).

Books (Published)

1. Bairagi, N.K. **Advanced Solid Mechanics**, Khanna Publishers, Delhi, 1985, pp. 1-250.

   [**see** Ch 1, Art 1.2, pp. 5-7 : Mathematical modelling for materials].

2. Bairagi, N.K. **Plate Analysis**, Khanna Publishers, Delhi, 1986, pp. 1-310.

   [**see** Ch 12, pp.283-306 : Plate analogy applied to the Selfing Theory of concrete mixes].

3. Bairagi, N.K. **Shell Analysis**, Khanna Publishers, Delhi, 1990, pp. 1-1303.

Technical Papers (Published)

1. Bairagi, N.K. and Kar, J.N. **Effect of Presetting on Mortar Strength**, *Cement and Concrete*, April-June, V-1, N-1, **1970**, pp. 25-29.

2. Bairagi, N.K. and Jhaveri, P.P. **Strength Variation of Composite Mixes by Pure Selfing Theory**, *Indian Concrete Journal,* March, V-53, N-3, **1977**, pp.87-89.

3. Bairagi, N.K. **Strength Variation of Composite Mixes by General Selfing Theory**, *Int. Conference on Materials of Construction for Developing Countries, Bangkok.* August 22-28, **1978** at AIT-Bangkok, Thailand (paper accepted, invited to read out the paper).

4. Bairagi, N.K. and Basavarajappa, G.H. **Use of Cylindrical Shells as Bridge Deck**, *Int. Conference on Computer Application in Civil Engineering,* October 23-25, **1979**, pp.109-113, *Uni. of Roorkee.*

5. Bairagi, N.K. Goyal, A.S. and Abdul Latheef, P.K. **Study of the Effect of Cumulative Curing on Strength of Concrete**, *Indian Concrete Journal,* January, **1989**, V-63, N-1, pp. 37-42.

6. Bairagi, N.K. Goyal, A.S. and Joshi, P.A. **Strength of Composite Mixes Using General Crossing Theory**, *Indian Concrete Journal,* December, **1989**, V-63, N-12, pp. 600-612.

7. Bairagi, N.K. Goyal, A.S. and Rao, M.R.**Selfing and Crossing Theories on Cumulatively Cured Strengths**, *Indian Concrete Journal*, November, **1990**, V-64, N-11, pp. 527-538.

8. Bairagi, N. K. and Ravande, K. **Mix Design Procedure for Recycled Aggregate Concrete**. *Construction and Building Materials (U. K).*, V-14, N-4, **1990**, pp. 78 -193.

9. Bairagi, N.K., Vidyadhara, H.S. and Ravande, Kishore **Mix Design Procedure for Recycled Aggregate Concrete**, *Construction and Building Materials (U.K),* December **1990**, V-4, N-4, pp. 188-193.

10. Bairagi, N. K. and Ravande, K. **Recycled Fine Aggregate—a Substitute for Natural Sand**. *Proceedings of International Conference on Recycling of Waste Material (Geneva, Switzerland)*, Jan. **1993**, pp. 11-12.

11. Bairagi, N. K., Ravande, K. and. Pareek, V. K. **Behaviour of Concrete with Different Proportions of Natural and Recycled Aggregates**. *Resources Conservation and Recycling (U. K).*, V-9, **1993**, pp. 109-126.

12. Bairagi, N.K. and Kandvinde, D.S. **Torsion of Concrete Beams Using Pre-Twisting Technique**. *Construction and Building Materials (U.K),* March **1993**, V-7, N-1, pp.45-51.

13. Bairagi, N.K., and Mohammed Ismail G. **Behaviour of Cylindrical Shells used as Bridge Decks**. *IASS (Madrid),* April **1994**, V-35,N-114, pp. 47-53.

14. Bairagi, N.K. and Mohammed Ismail G. **Effect of Vehicular Loading on Circular Cylindrical Shells Used as Bridge Decks**. *Indian National Group of IABSE,* September **1994**, V-24, N-3, pp. 1-12.

15. Bairagi, N.K. and Mahmood M. El-Frnsawy. **Strength of Concrete Using Selfing Theory** *National Conference on Civil Engg. Material and Structures, Osmania Univ. (Hyderabad)* January 19-21, **1995**, pp.149-155.

16. Bairagi, N.K. and Mahmood M. El-Frnsawy. **Interfacial Bond between Two Preset Concrete Layers**. *National Conference on Civil Engg. Material and Structures, Osmania Univ. (Hyderabad)* January 19-21, **1995**, pp.109-116.

17. Bairagi, N.K. and Pise, P.T. **Torsion of Concrete Beams Using Post-Twisting Technique.** *Construction and Building Materials (U.K),* **1995**, pp. 135-139.

18. Bairagi, N.K. and Babu Kurian. **Torsion of Concrete Beams Using Pretwisting Technique**. *National Conference on Civil Engg. Material and Structures, Osmania Univ. (Hyderabad)* January 19-21, **1995**, pp. 241-246.

19. Bairagi, N. K., Dubal, N. S. **Effect of Thermal Cycles on Concrete**. *International Journal of Structures,* V-16, V-1, **1996**, pp. 23-27.

20. Modhera, C.D., Ravi Akhilesh Kumar and Bairagi, N.K. **Behaviour of Glass Fibre Reinforced Concrete in Cumulative Curing Condition**. *22nd Conference on 'Our World in Concrete and Structures' August 25-27, 1997, Singapore,* V -XVI, pp. 253-258.

21. Bairagi, N.K. and Mahmood El-Frnsawy. **Selfing Concept—its Application to Interfacial Strength**. *Proceedings of Seminar on Design Management Construction and Computer, B.E. College (Deemed Univ.),* January 25-26, **1997**. Technical Session III, pp.1-7.

22. Bairagi, N.K. and Mahmood M. El-Frnsawy **Selfing of Blended Mixes —Application and Interpretations**. *First International Conference on Civil Engineering, 24-26 March 1998, Helwan University, Cairo, Egypt*, pp.13-23.

23. Bairagi, N. K. and Modhera, C. D. **Study on Setting Time of Concrete Using Penetration Resistance Method**. *International Journal of Structures.*, V-18, N-2, July- Dec. **1998**, pp. 7–8.

24. Bairagi, N.K. and Mahmood M. El-Frnsawy. **Spring Function Concept—Its Application in Structural Spring Systems**. *First International Conference on Civil Engineering, 24-26 March 1998, Helwan University, Cairo, Egypt,* pp. 72-78.

25. Ravande, Kishore, Bairagi, N.K. and Kumar, P. **Strength Characteristics of Concrete Made with Recycled Fine Aggregate and Crushed Stone Fines.** *The 15th International Conference on Solid Waste Technology and Management., 12-15th December, 1999, Philadelphia, USA.*

26. Bairagi, N.K. and Modhera, C.D. **Shear Strength of Fibre Reinforced Concrete**. *Indian Concrete Institute,* Jan-Mar. **2001**, pp. 7-52.

Index

Symbols

A

B

C

D

E

F

G

H

N

O

P

Q

R

S

T

U

V

W

Z